Bioinformatics Applications Based On Machine Learning

Bioinformatics Applications Based On Machine Learning

Editors

Pablo Chamoso
Sara Rodríguez González
Mohd Saberi Mohamad
Alfonso González-Briones

MDPI • Basel • Beijing • Wuhan • Barcelona • Belgrade • Manchester • Tokyo • Cluj • Tianjin

Editors
Pablo Chamoso
University of Salamanca
Spain

Sara Rodríguez González
University of Salamanca
Spain

Mohd Saberi Mohamad
United Arab Emirates University
United Arab Emirates

Alfonso González-Briones
Complutense University of Madrid
Spain

Editorial Office
MDPI
St. Alban-Anlage 66
4052 Basel, Switzerland

This is a reprint of articles from the Special Issue published online in the open access journal *Processes* (ISSN 2227-9717) (available at: https://www.mdpi.com/journal/processes/special_issues/Bioinformatics_Application).

For citation purposes, cite each article independently as indicated on the article page online and as indicated below:

LastName, A.A.; LastName, B.B.; LastName, C.C. Article Title. *Journal Name* **Year**, *Volume Number*, Page Range.

ISBN 978-3-0365-0760-6 (Hbk)
ISBN 978-3-0365-0761-3 (PDF)

Cover image courtesy of unsplash.com user Pietro Jeng.

Contents

About the Editors

Pablo Chamoso is Assistant Professor at the University of Salamanca. He holds a PhD in Computer Engineering (2017) as well as qualifications of Technical Engineer in Computer Systems, Computer Engineer (with i3 award for the best final project of Castilla y León), Master in Electronic Commerce, Master in Intelligent Systems, and Master in Information Technology Management.

He has been a member of the research group BISITE since early 2011, participating in the technical development, analysis, and research of multiple international and national projects in areas such as smart cities, robotics, visual analytics, medical applications, and artificial intelligence.

He has been a guest lecturer at universities in Brazil, Portugal, Italy, and Japan. He is co-author of numerous publications in high-impact scientific journals and serves as editor or reviewer for several JCR-indexed journals. He has been a member of the organization and scientific committee of numerous international scientific conferences. His research career has been recognized and awarded with one of the Grants of the State Program Juan de la Cierva in the area "ICT—Information and Communication Technologies" in 2018.

Sara Rodríguez González is Associate Professor at the Department of Computer Science and Automatics, University of Salamanca, and researcher in the BISITE research group (http://bisite.usal.es). She pursued her PhD studies at this University, obtaining an Engineering in Computer Sciences degree in 2007 and being awarded a PhD in Computer Science in 2010. She has participated as a co-author in more than 100 papers published in recognized international conferences and symposiums. She has worked on 50 research projects and around 70 research contracts. She has been active in the organization of numerous international scientific congresses (PAAMS, CEDI, FUSION, DCAI, etc.).

Mohd Saberi Mohamad is Professor of Bioinformatics and Artificial Intelligence at the Department of Genetics and Genomics, the College of Medicine and Health Sciences, UAE University. His research areas are bioinformatics, artificial intelligence, data science, and computational biology. He has received the 2018 Malaysia's Research Star Award in the category of young researchers by the Malaysia Ministry of Higher Education. He has been the project leader for 19 research grants and a project member for 20 research grants. He has also published 282 articles published in international refereed journals, international conference proceedings, and as book chapters in addition to being involved in the production of 14 books (research books, edited books, and original books). Prior to joining UAUE, he has been Director of the Institute for Artificial Intelligence and Big Data, founder of the Department of Data Science, and head of the Artificial Intelligence and Bioinformatics Research Group. In addition, he has also served as head and member of 65 academic and research committees at the faculty, university, national, and international levels. He also has been appointed as an advisory board for Artificial Intelligence Research Institute and IoT Digital Innovation Hub in Europe. He has served as chairman in 7 appointments and member of the committee for 39 international conferences in Europe, ASEAN, Japan, US, Australia, China, and Malaysia. He has served on the accreditation panel of Malaysia Qualification Agency (MQA), which reviews academic programs in the field of bioinformatics, since 2013.

Alfonso González-Briones was awarded his PhD in Computer Engineering from the University of Salamanca in 2018, and his thesis obtained second place in the 1st SENSORS+CIRTI Award for the best national thesis in Smart Cities (CAEPIA 2018). At the same university, he obtained his Bachelor of Technical Engineer in Computer Engineering (2012), Degree in Computer Engineering (2013), and Master in Intelligent Systems (2014). Currently, he is a "Juan De La Cierva" Postdoc at GRASIA Research Group in the University Complutense of Madrid at Software Engineering and Artificial Intelligence Department. Alfonso González Briones has published more than 20 articles in journals, more than 40 articles in books and international congresses, and has participated in 8 international research projects. He is also a member of the scientific committee of the *Advances in Distributed Computing and Artificial Intelligence Journal (ADCAIJ)* and *British Journal of Applied Science & Technology (BJAST)* and is also a regular reviewer of renowned journals, namely published by IEEE, Springer, Elsevier, or MDPI, among others. He has participated as chair and member of the technical committee of prestigious international congresses (AIPES, HAIS, FODERTICS, PAAMS, KDIR).

Preface to "Bioinformatics Applications Based On Machine Learning"

Research in the area of bioinformatics has always been one of the most active lines of research in the scientific community. However, it has recently gained even more interest thanks to advances in the information technology (IT) sector, including the increased processing capacities of computers, which allow processing large volumes of data and analyzing them with techniques such as machine learning.

Thanks to these advances, new applications appear in the area of bioinformatics. In them, the results obtained generally improve those of previous applications that do not use these computation techniques.

This book presents papers that have been accepted for the Special Issue "Bioinformatics Applications Based On Machine Learning" of the journal Processes, where authors were encouraged to submit their original research dealing with new machine learning algorithms, distributed machine learning systems, new applications in bioinformatics, healthcare applications, bioimaging, next-generation sequencing, data and software integration, visualization of biological systems and networks, high-throughput data analysis (transcriptomics, proteomics, etc.), comparison and alignment methods, and other related topics.

After several rounds of review, 10 research articles and 1 review (entitled "A Review of Computational Methods for Clustering Genes with Similar Biological Functions") were accepted for publication in the Special Issue and are included in this book.

The research articles include the use of a wide variety of IT techniques such as convolutional neural networks, gradient boosting, multilayer bi-directional LSTM, particle swarm optimization or harmony search, among others, applied to domains such as body part detection in images and video, diabetes, or the study of *Arabidopsis thaliana* or *Saccharomyces cerevisiae*, among others.

Finally, as the Guest Editors, we would like to take this opportunity to thank all the contributing authors and the reviewers for their hard and highly valuable work. In addition, we acknowledge the funding support for the project "Intelligent and sustainable mobility supported by multi-agent systems and edge computing (InEDGEMobility): Towards Sustainable Intelligent Mobility: Block-chain-based framework for IoT Security", Reference: RTI2018-095390-B-C32, financed by the Spanish Ministry of Science, Innovation and Universities (MCIU), the State Research Agency (AEI) and the European Regional Development Fund (FEDER), and last but not least, the MDPI staff for their hard work, which was essential for the success of this book.

Pablo Chamoso, Sara Rodríguez González, Mohd Saberi Mohamad, Alfonso González-Briones

Editors

Article

Population-Based Parameter Identification for Dynamical Models of Biological Networks with an Application to *Saccharomyces cerevisiae*

Ewelina Weglarz-Tomczak [1,*,†], Jakub M. Tomczak [2,*,†], Agoston E. Eiben [2] and Stanley Brul [1]

[1] Swammerdam Institute for Life Sciences, Faculty of Science, University of Amsterdam, 1090 GE Amsterdam, The Netherlands; s.brul@uva.nl
[2] Department of Computer Science, Faculty of Science, Vrije Universiteit Amsterdam, 1081 HV Amsterdam, The Netherlands; a.e.eiben@vu.nl
* Correspondence: ewelina.weglarz.tomczak@gmail.com (E.W.-T.); jmk.tomczak@gmail.com (J.M.T.)
† These authors contributed equally to this work.

Abstract: One of the central elements in systems biology is the interaction between mathematical modeling and measured quantities. Typically, biological phenomena are represented as dynamical systems, and they are further analyzed and comprehended by identifying model parameters using experimental data. However, all model parameters cannot be found by gradient-based optimization methods by fitting the model to the experimental data due to the non-differentiable character of the problem. Here, we present POPI4SB, a Python-based framework for population-based parameter identification of dynamic models in systems biology. The code is built on top of PySCeS that provides an engine to run dynamic simulations. The idea behind the methodology is to provide a set of derivative-free optimization methods that utilize a population of candidate solutions to find a better solution iteratively. Additionally, we propose two surrogate-assisted population-based methods, namely, a combination of a k-nearest-neighbor regressor with the Reversible Differential Evolution and the Evolution of Distribution Algorithm, that speeds up convergence. We present the optimization framework on the example of the well-studied glycolytic pathway in *Saccharomyces cerevisiae*.

Keywords: dynamic models; evolutionary computing; derivative-free optimization; metabolism; glycolysis; yeast

Citation: Weglarz-Tomczak, E.; Tomczak, J.M.; Eiben, A.E.; Brul, S. Population-Based Parameter Identification for Dynamical Models of Biological Networks with an Application to *Saccharomyces cerevisiae*. *Processes* **2021**, *9*, 98. https://doi.org/10.3390/pr9010098

Received: 14 December 2020
Accepted: 30 December 2020
Published: 5 January 2021

Publisher's Note: MDPI stays neutral with regard to jurisdictional claims in published maps and institutional affiliations.

1. Introduction

Mathematical models in systems biology are mostly represented by ordinary differential equations (ODEs). They provide a representation of the information obtained from experimental observations about the structure and function of a particular biological network [1,2]. The integral component of ODEs is parameters that correspond to the kinetic characteristics of a reaction catalyzed by a specific enzyme in particular conditions. Typically, the parameters are identified by fitting the model to experimental data or are measured for individual reactions separately. Once parameter values are determined, dynamic models could be used to confirm hypotheses, draw predictions and find such (time-varying) stimulation conditions that result in a particular desired behavior of a system [2–4]. However, the problem of fitting a dynamical model to experimental data is non-differentiable, thus, derivative-free optimization methods should be used instead of gradient-based or higher-order optimizers [5,6].

Here, we present a framework that implements a set of population-based optimization methods to identify parameters in a dynamic model of a biological network of interest, from limited available experimental data. In other words, the presented framework allows finding parameter values of a dynamical model while only selected quantities are observed. This could drastically decrease the time of fitting separate reactions to data and improve estimation quality because all reactions are considered as a whole, thus, it takes into accou[nt] interaction among reactions. The implementation of the approach is a stand-alone Pyth[on] program. It utilizes PySCeS (Python Simulator for Cellular System) [7], a modeling tool f[or] formulating dynamical models of biological networks and running simulations by solvi[ng] ODEs numerically. Our framework loads a model developed using PySCeS or from t[he] JWS database [8] together with experimental data, and outputs parameter values for whi[ch] a difference between the experimental data and the simulation is smallest. Moreover, t[he] framework allows adding new optimizers to a single file, without the necessity of changi[ng] any other parts of the program. Please see *Supplementary Data* for details. We refer to t[he] framework as POPI4SB, see its schematic representation in Figure 1.

In this study, we chose *glycolysis* that is a crucial metabolic pathway and its upre[gu]lation is correlated with diseases like cancer [9,10]. Nearly all living organisms carry o[ut] glycolysis as a part of cellular metabolism. One of the most intensively studied organis[ms] in the context of, among others, glycolysis is *Saccharomyces cerevisiae* species, also known [as] baker's yeast [11–15]. We applied our optimization framework to a model of glycolysis [in] yeast proposed in [16]. This model contains lumped reactions of the glycolytic pathw[ay] and includes production of glycerol, fermentation to ethanol and exchange of acetaldehy[de] between the cells, and trapping of acetaldehyde by cyanide.

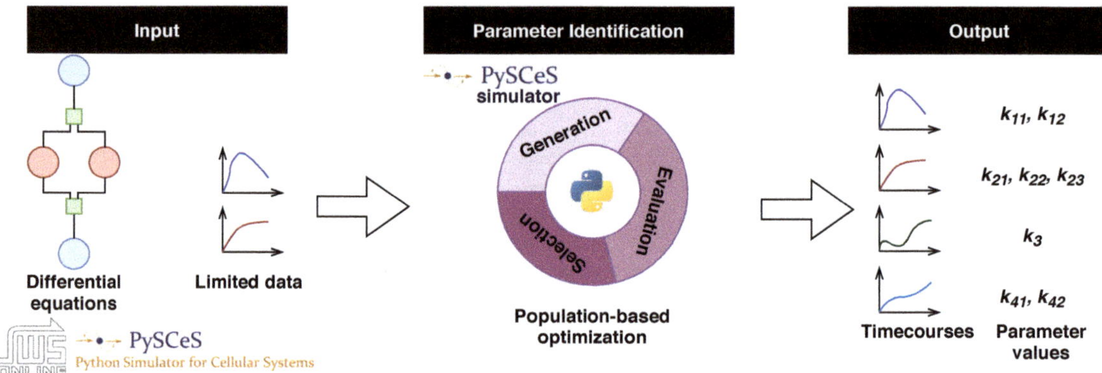

Figure 1. A schematic representation of our framework. A dynamic model in the PySCeS format and experimental data are inputs to the program. The core component is the parameter identification with population-based optimization methods. Eventually, parameters values are returned, for with the lowest error (i.e., the difference between simulated data and experimental data) was achieved.

The contribution of the paper is threefold:

- We provide a population-based optimization framework for parameter identificati[on] and showcase its performance on the example of the glycolysis of *Saccharomy[ces] cerevisiae*, one of the most studied species in biology.
- We analyze the performance of the population-based optimization framework in t[he] considered problem and indicate its high potential for future research.
- We extend the Python framework PySCeS [7] by implementing the population-bas[ed] optimization methods (four methods known in the literature, and two new metho[ds] in Python. The code for the methods together with the experiments is available onli[ne] https://github.com/jmtomczak/popi4sb.

2. Materials and Methods

2.1. Derivative-Free Optimization

We consider an optimization problem of a function $f : \mathbb{X} \to \mathbb{R}$, where $\mathbb{X} \subseteq \mathbb{R}^D$ is the search space. In this paper we focus on the minimization problem, namely:

$$\mathbf{x}^* = \arg \min_{\mathbf{x} \in \mathbb{X}} f(\mathbf{x}; \mathcal{D}), \tag{1}$$

where $\mathcal{D}$ denotes observed data.

Further, we assume that the analytical form of the function f is unknown or cannot be used to calculate derivatives, however, we can query it through a simulation or experimental measurements. Problems of this sort are known as *derivative-free* or *black-box* (In general, a *black-box* problem means that a formal description of a problem is unknown, however, very often non-differentiable problems with known mathematical representation (e.g., differential equations) are treated as black-box) optimization problems [5,17]. Additionally, we consider a bounded search space, i.e., we include inequality constraints for all dimensions in the following form: $l_d \leq x_d \leq u_d$, where $l_d, u_d \in \mathbb{R}$ and $l_d < u_d$, for $d = 1, 2, \ldots, D$.

2.2. Population-Based Optimization Methods

One group of widely-used methods for derivative-free optimization problems is population-based optimization algorithms. The idea behind these methods is to use a *population* of *individuals*, i.e., a collection of candidate solutions $\mathcal{X} = \{\mathbf{x}_1, \ldots, \mathbf{x}_N\}$, instead of a single individual in the iterative manner. The premise of utilizing the population over a single candidate solution is to obtain better exploration of the search space and exploiting potential local optima [18,19].

In the essence, every population-based algorithm consists of three following steps that utilize a procedure for generating new individuals G, and a selection procedure S, that is:

(Init) Initialize $\mathcal{X} = \{\mathbf{x}_1, \ldots, \mathbf{x}_N\}$ and evaluate all individuals $\mathcal{F}_x = \{f_n : f_n = f(x_n), x_n \in \mathcal{X}\}$.

(Generation) Generate new candidate solutions using the current population, $\mathcal{C} = G(\mathcal{X}, \mathcal{F}_x)$.

(Evaluation) Evaluate all candidates solutions:

$$\mathcal{F}_c = \{f_n : f_n = f(x_n), x_n \in \mathcal{C}\}.$$

(Selection) Select a new population using the candidate solutions and the old population

$$\mathcal{X} := S(\mathcal{X}, \mathcal{F}_x, \mathcal{C}, \mathcal{F}_c).$$

Go to **Generate** or terminate.

An exemplary population-based optimization approach is depicted in Figure 2.

In general, the population-based optimization methods are favorable over standard derivative-free optimization (DFO) algorithms in problems when querying the objective function is relatively cheap. Their computational complexity depends mainly on the population size, i.e., it is linear with respect to the size of the population N. Other DFO methods are typically more expensive. Bayesian Optimization, for instance, is known to give a good performance, but its complexity typically scales cubically with respect to the number of queries [20]. Here, we take advantage of the very low execution time of running a simulator (the glycolysis model) and propose to use the population-based methods for the parameter identification task.

There are a plethora of population-based DFO algorithms [5,6,18,21,22], however, our goal is to verify whether this approach, in general, could be successfully used in the considered task. Therefore, we decide to choose four instances of a group of methods that are easy-to-use and are proven to work well in practice: evolutionary strategies (ES), differential evolution (DE), estimation of distribution algorithms (EDA), and recently

proposed reversible differential evolution (RevDE). Moreover, we propose to enhance ED[
and RevDE with a surrogate model to allow better exploration and speed up calculatior

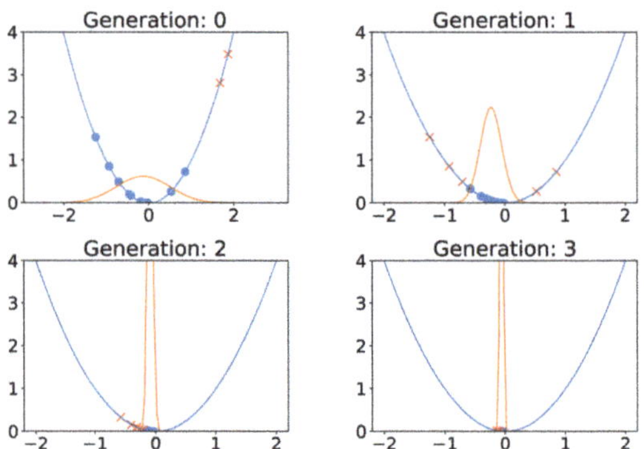

Figure 2. An illustration of a population-based optimization of a quadratic function (blue sol[
line) using the Estimation of Distribution Algorithm. At each generation a population is selecte
(blue nodes) and weakest individuals are discarded (red crosses). New candidate solutions a[
generated by sampling from the normal distribution fit to the previous population (orange solid line

2.2.1. Evolutionary Strategies (ES)

Evolutionary strategies can be seen as a specialization of evolutionary algorithms wit
very specific choices of G and S. The core of ES is to formulate G using the multivaria
Gaussian distribution. Here, we follow the widely-used $(1 + 1)$-ES that generates a ne
candidate using the Gaussian mutation parameterized by $\sigma > 0$, namely:

$$\mathbf{x}' = \mathbf{x} + \sigma \cdot \varepsilon,$$

where $\varepsilon \sim \mathcal{N}(0, I)$, and $\mathcal{N}(0, I)$ denotes the Gaussian distribution with zero mean and th
identity covariance matrix I. Next, if the fitness value of $\mathbf{x}'$ is smaller than the value c
fitness function of $\mathbf{x}$, the new candidate is accepted and the old one is discarded.

The crucial element of this approach is determining the value of σ. In order to ove
come possibly time-consuming hyperparameter search, the following adaptive procedu[
is proposed [21]:

$$\sigma := \begin{cases} \sigma \cdot c & \text{if } p_s < 1/5, \\ \sigma/c & \text{if } p_s > 1/5, \\ \sigma & \text{if } p_s = 1/5. \end{cases}$$

where p_s is the number of accepted individuals of the offspring divided by the populatic
size N, and c is equal 0.817 following the recommendation in [23].

2.2.2. Differential Evolution (DE)

Differential evolution is another population-based method that is loosely based on th
Nelder-Mead method [24,25]. A new candidate is generated by randomly picking a trip[
from the population, $(\mathbf{x}_i, \mathbf{x}_j, \mathbf{x}_k) \in \mathcal{X}$, and then $\mathbf{x}_i$ is perturbed by adding a scaled differenc
between $\mathbf{x}_j$ and $\mathbf{x}_k$, that is:

$$\mathbf{y} = \mathbf{x}_i + F(\mathbf{x}_j - \mathbf{x}_k),$$

where $F \in (0, 2]$ is the scaling factor. This operation could be seen as an adaptive *mutatic
operator* that is widely known as *differential mutation* [25].

Further, the authors of [24] proposed to sample a binary mask $\mathbf{m} \in \{0,1\}^D$ according to the Bernoulli distribution with probability $p = P(m_d = 1)$ shared across all D dimensions, and calculate the final candidate according to the following formula:

$$\mathbf{v} = \mathbf{m} \odot \mathbf{y} + (1 - \mathbf{m}) \odot \mathbf{x}_i, \tag{5}$$

where $\odot$ denotes the element-wise multiplication. In the evolutionary computation literature this operation is known as *uniform crossover operator* [18]. In this paper, we fix $p = 0.9$ following general recommendations in literature [26] and use the uniform crossover in all methods.

The last component of a population-based method is a selection mechanism. There are multiple variants of selection [18], however, here we use the "survival of the fittest" approach, i.e., we combine the old population with the new one and select N candidates with highest fitness values, i.e., the deterministic $(\mu + \lambda)$ selection.

This variant of DE is referred to as "DE/*rand*/1/*bin*", where *rand* stands for randomly selecting a base vector, *1* is for adding a single perturbation and *bin* denotes the uniform crossover. Sometimes it is called *classic DE* [25].

2.2.3. Reversible Differential Evolution (RevDE)

The mutation operator in DE perturbs candidates using other individuals in the population to generate a single new candidate. As a result, having too small population could limit exploration of the search space. In order to overcome this issue, a modification of DE was proposed that utilized all three individuals to generate three new points in the following manner [27]:

$$\begin{aligned}
\mathbf{y}_1 &= x_i + F(\mathbf{x}_j - \mathbf{x}_k) \\
\mathbf{y}_2 &= x_j + F(\mathbf{x}_k - \mathbf{y}_1) \\
\mathbf{y}_3 &= x_k + F(\mathbf{y}_1 - \mathbf{y}_2).
\end{aligned} \tag{6}$$

New candidates $\mathbf{y}_1$ and $\mathbf{y}_2$ could be further used to calculate perturbations using points outside the population. This approach does not follow a typical construction of an EA where only evaluated candidates are mutated. Further, we can express (6) as a linear transformation using matrix notation by introducing matrices as follows:

$$\begin{bmatrix} \mathbf{y}_1 \\ \mathbf{y}_2 \\ \mathbf{y}_3 \end{bmatrix} = \underbrace{\begin{bmatrix} 1 & F & -F \\ -F & 1 - F^2 & F + F^2 \\ F + F^2 & -F + F^2 + F^3 & 1 - 2F^2 - F^3 \end{bmatrix}}_{=\mathbf{R}} \begin{bmatrix} \mathbf{x}_1 \\ \mathbf{x}_2 \\ \mathbf{x}_3 \end{bmatrix}. \tag{7}$$

In order to obtain the matrix $\mathbf{R}$, we need to plug $\mathbf{y}_1$ to the second and third equation in (6), and then $\mathbf{y}_2$ to the last equation in (6). As a result, we obtain $M = 3N$ new candidate solutions. This version of DE is called *Reversible Differential Evolution*, because the linear transformation $\mathbf{R}$ is reversible [27].

2.2.4. Estimation of Distribution Algorithms (EDA)

Most of the population-based optimization methods aim at finding a solution and the information about the distribution of the search space and the fitness function is represented implicitly by the population. However, this distribution could be modeled explicitly using a probabilistic model [19]. These methods have become known as the estimation of distribution algorithms [28–30].

The key difference between EDA and EA is the generation step. While an EA uses evolutionary operators like mutation and cross-over to generate new candidate solutions, EDA fits a probabilistic model to the population, and then new individuals are sampled from this model.

Therefore, fitting a distribution to the population is the crucial part of an ED□ There are various probabilistic models that could be used for this purpose. Here, we p□ pose to fit the multivariate Gaussian distribution $\mathcal{N}(\bar{\ },\Sigma)$ to the population $\mathcal{X}$. For t□ purpose, we can use the empirical mean and the empirical covariance matrix:

$$\hat{\mu} = \frac{1}{N} \sum_{n=1}^{N} \mathbf{x}_n,$$

and

$$\hat{\Sigma} = \frac{1}{N} \sum_{n=1}^{N} (\mathbf{x}_n - \hat{\mu})(\mathbf{x}_n - \hat{\mu})^{\top}.$$

An efficient manner of sampling new candidates is to first calculate the Choles□ decomposition of the covariance matrix, $\hat{\Sigma} = \mathbf{L}\mathbf{L}^{\top}$, where $\mathbf{L}$ is the lower-triangular mat□ and then computing:

$$\mathbf{x}' = \hat{\mu} + \mathbf{L}\varepsilon, \tag{□}$$

where $\varepsilon \sim \mathcal{N}(0, \mathbf{I})$. The Equation (10) is repeated M times to generate a new set of candid□ solutions. Here, we set M to the size of the population, i.e., $M = N$. Once new candid□ solutions are generated, the selection mechanism is applied. In this paper, we use the sa□ selection procedure as the one used for DE.

2.2.5. Population-Based Methods with Surrogate Models (RevDE+ & EDA+)

Surrogate models: A possible drawback of population-based methods is the nec□ sity of evaluating large populations that, even though we assume a low time cost pe□ single evaluation, could significantly slow down the whole optimization process. To ov□ come this issue, a surrogate model could be used to partially replace querying the fitn□ function [31]. The surrogate model is either a probabilistic model or a machine learni□ model (e.g., a neural network) that gathers previously evaluated populations and allo□ to mimic the behavior of the fitness function. While applying the surrogate model, i□ assumed that its utilization cost (e.g., training) is lower or even significantly lower than t□ computational cost of running the simulator.

There are multiple possible surrogate models, however, non-parametric mod□ e.g., Gaussian processes [20], are preferable, because they do not suffer from catastrophic f□ getting (i.e., overfitting to the last population and forgetting first populations). Here, we c□ sider another non-parametric model, namely, K-Nearest-Neighbor (K-NN) regressi□ model that stores all previously seen individuals with evaluations, and the prediction o□ new candidate solution is an average over K (e.g., $K = 3$) closest previously seen indivi□ als. Current implementations of the K-NN regressor provide efficient search procedu□ that result in the computational complexity better than $N \cdot D$, e.g., using KD-trees results□ $O(D \log N)$. This computational complexity is significantly better than the computatio□ complexity of Gaussian processes, $O(N^3)$.

RevDE+: In the RevDE approach, we generate $3N$ new candidate solutions and□ of them are further evaluated. However, this introduces an extra computational cost□ running the simulator. This issue could be alleviated by using the K-NN regressor□ approximate the fitness values of the new candidates. Further, we can select N m□ promising points. We refer to this approach as RevDE+.

EDA+: The outlined procedure of EDA produces M new candidate solutions a□ to keep a similar computational cost as ES and DE, we set M to N. However, this cou□ significantly limit the potential of modeling a search space, because sampling in hig□ dimensional search spaces requires a significantly large number of points. A potent□ solution to this problem could be the application of the K-NN regressor to quickly ver□ the L new points. As long as the time cost of providing the approximated value of □ fitness function is lower than the running time of the simulator, we can afford to ta□ $L > N$ (e.g., $L = 5N$). We refer to this approach as EDA+.

2.3. The Model of Glycolysis in Saccharomyces Cerevisiae

Introduction: As an example, we chose *glycolysis* that is a crucial metabolic pathway and its upregulation is correlated with diseases like cancer [9,10]. Nearly all living organisms carry out glycolysis as a part of cellular metabolism. A glycolytic path that consists of a series of reactions breaks down glucose into two three-carbon compounds and extracts energy for cellular metabolism. Therefore, glycolysis is at the heart of classical biochemistry and, as such, it is very well described. One of the most intensively studied organisms in the context of, among others, glycolysis is *Saccharomyces cerevisiae* species, also known as baker's yeast [11–15]. Whereas, the dynamic model of glycolysis in *Saccharomyces cerevisiae* is of big interest in systems biology dynamic modeling literature [16,32–35].

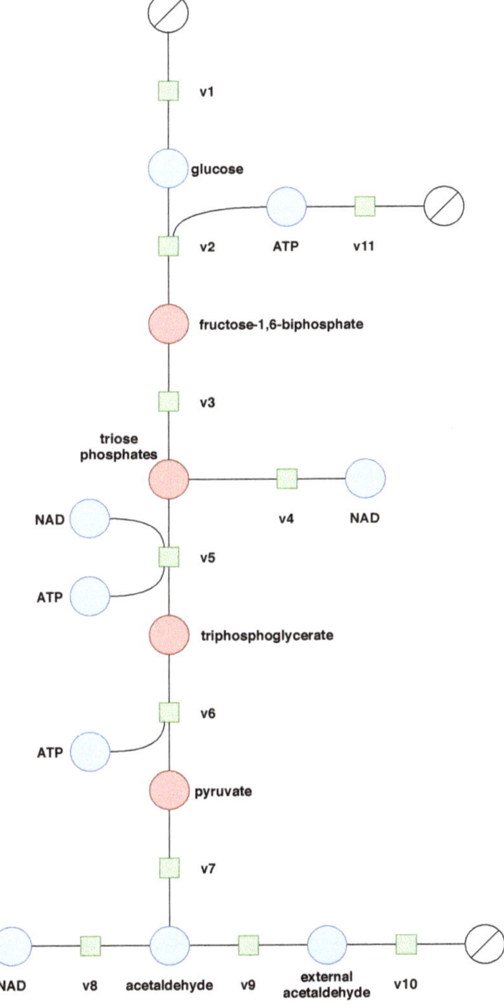

Figure 3. The glycolysis process in the yeast *Saccharomyces cerevisiae* proposed in [16]. There are 11 reactions governing the process with 18 parameters in total, and 9 metabolites. Blue circles depict observable metabolites, red circles denote unobservable metabolites, and green squares represent reactions. A white circle with a diagonal line corresponds to a sink. The model is taken from the JWS database [8].

We applied our optimization framework to a model of glycolysis in yeast propose[...] in [16], see Figure 3, that suffices to present the essence of our framework. This model co[...] tains lumped reactions of the glycolytic pathway and includes the production of glycer[...] fermentation to ethanol, and exchange of acetaldehyde between the cells, and trapping [...] acetaldehyde by cyanide.

A system of Ordinary Differential Equations: In the considered model of the glyc[...] ysis we distinguish the following metabolites: glycolysis (glu), fructose-1,6-bisphospha[...] (fru), triosephosphates ($triop$), triphosphoglycerate (tp), pyruvate (pyr), acetaldehyde (a[...] external acetaldehyde (ace).

Following the same assumptions as in [16] (i.e., a homogeneous distribution of th[...] metabolites in the intracellular and in the extracellular solution), the system of ordina[...] differential equations of the glycolysis model in *Saccharomyces cerevisiae* is the following [3[...]

$$\dot{glu} = v_1 - v_2 \tag{1}$$
$$\dot{fru} = v_2 - v_3 \tag{1}$$
$$\dot{triop} = 2v_3 - v_4 - v_5 \tag{1}$$
$$\dot{tp} = v_5 - v_6 \tag{1}$$
$$\dot{pyr} = v_6 - v_7 \tag{1}$$
$$\dot{ac} = v_7 - v_8 - v_9 \tag{1}$$
$$\dot{ace} = 0.1v_9 - v_{10} \tag{1}$$
$$\dot{atp} = -2v_2 + v_5 + v_6 - v_{11} \tag{1}$$
$$\dot{nad} = v_4 - v_5 - v_8 \tag{1}$$

with the rate equations:

$$v_1 = k_0 \tag{2}$$
$$v_2 = \frac{k_1 \cdot glu \cdot at}{1 + (at/k_i)^n} \tag{2}$$
$$v_3 = k_2 \cdot fru \tag{2}$$
$$v_4 = \frac{k_{31} \cdot k_{32} \cdot triop \cdot nad\,A - k_{33} \cdot k_{34} \cdot tp \cdot atp\,N}{k_{33} \cdot N + k_{32} \cdot A} \tag{2}$$
$$v_5 = k_4 \cdot tp \cdot A \tag{2}$$
$$v_6 = k_5 \cdot pyr \tag{2}$$
$$v_7 = k_6 \cdot ac \cdot nad \tag{2}$$
$$v_8 = k_7 \cdot atp \tag{2}$$
$$v_9 = k_8 \cdot triop \cdot nad \tag{2}$$
$$v_{10} = k_9 \cdot ace \tag{2}$$
$$v_{11} = k_7 \cdot atp \tag{3}$$

where $A = (a_{tot} - atp)$ and $N = (n_{tot} - nad)$.

Initial conditions

The initial conditions are the following:

$$\begin{aligned}
atp &= 2.0 & nad &= 0.6 & glu &= 5.0 \\
fru &= 5.0 & triop &= 0.6 & tp &= 0.7 \\
pyr &= 8.0 & ac &= 0.08 & ace &= 0.02\,.
\end{aligned}$$

Real parameter values

The real values of the parameters are the following [36]:

$$a_{tot} = 4 \in [0, 10] \qquad k_0 = 0 \in [0, 10] \qquad k_1 = 550 \in [550, 600]$$
$$k_2 = 9.8 \in [0, 10] \qquad k_{31} = 323.8 \in [300, 350] \qquad k_{32} = 76411.1 \in [76,400, 76,450]$$
$$k_{33} = 57823.1 \in [57800, 57850] \qquad k_{34} = 23.7 \in [20, 50] \qquad k_4 = 80 \in [80, 100]$$
$$k_5 = 9.7 \in [0, 10] \qquad k_6 = 2000 \in [2000, 2050] \qquad k_7 = 28.0 \in [20, 50]$$
$$k_8 = 85.7 \in [80, 100] \qquad k_9 = 0 \in [0, 10] \qquad k_{10} = 375 \in [350, 400]$$
$$k_i = 1 \in [0, 10] \qquad n = 4 \in [0, 10] \qquad n_{tot} = 1 \in [0, 10]$$

where we indicate the set of possible values of the parameters in the square brackets.

We note that for the sake of our experiments, we set k_0 to 0 (originally: $k_0 = 50$ [36]) in order to forbid a constant injection of *glu*, and k_9 to 0 (originally: $k_9 = 80$ [36]) in order to avoid oscillatory behavior of the system.

3. Experimental Setup

The experiments have been carried *in silico* in which the performance of the selected algorithms has been evaluated.

3.1. Implementation

POPI4SB is implemented in Python, and utilizes PySCeS for running simulations. The code for carrying out experiments is available online: https://github.com/jmtomczak/popi4sb. The list of requirements is provided therein.

3.2. Parameter Identification & the Fitness Function

We consider the glycolysis process in yeast as a biochemical system with inputs and outputs (see Figure 3). The input to the system is glucose (*glu*), and the outputs are ATP (*atp*), NAD (*nad*), acetaldehyde (*ac*), and external acetaldehyde (*ace*). The other metabolites, i.e., triose phosphates (*triop*), pyruvate (*pyr*), fructose-1,6-biphosphate (*fru*) and triphosphoglycerate (*tp*) are considered to be unobserved quantities. The system is governed by 11 reactions with 18 parameters in total (see Appendix for details). Each reaction is represented by an ordinary differential equation that is known. We assume that we have inputs and outputs, namely, i.e., *glu*, *atp*, *nad*, *ac*, and *ace*, and each quantity is represented as a timecourse of length T. We denote these measurements by

$$\mathcal{D} = \{glu, atp, nad, ac, ace\}.$$

Further, following the nomenclature presented in [37], we consider the system of differential equations representing the glycolysis process as the **simulator** that for given values of parameters and initial conditions provides timecourses of all metabolites. Then, we can denote parameters by $\mathbf{x}$ and the simulator by sim : $\mathcal{X} \rightarrow \mathbb{R}^{9 \times T}$, i.e., sim takes parameters $\mathbf{x}$ and simulates timcourses of length T for all 9 metabolites, including *glu*, *atp*, *nad*, *ac*, *ace*. In order to calculate the objective (or the fitness) of the parameter values, we use the following function:

$$f(\mathbf{x}; \mathcal{D}) = \sum_{i=1}^{5} \frac{1}{\gamma \cdot T} \sum_{t=1}^{T} \|\mathbf{y}_{i,t} - \text{sim}_{i,t}(\mathbf{x})\|_2^2, \tag{31}$$

where $\mathbf{y}_{i,t}$ corresponds to one of the five observed metabolites at the t-th time step, and $\text{sim}_{i,t}(\mathbf{x})$ is the corresponding synthetically generated signal given by the simulator with parameters $\mathbf{x}$, $\gamma > 0$ specifies the strength of penalizing a mistake. Notice that this is the (unnormalized) logarithm of the product of Gaussian distributions with means given by $\text{sim}(\mathbf{x})$ and the diagonal covariance matrix with shared variance γ.

3.3. Simulated Data

In the experiments, we assume that *glu*, *atp*, *nad*, *ac*, and *ace* are observed. We generate the observed metabolites by running the simulator with the real parameter values. To mimic real measurements that are typically noisy, we add a Gaussian noise with zero mean and the standard deviation equal 3% of a generated value of a metabolite at a given time step.

Adding noise prohibits finding a solution (i.e., values of parameters) that achieves er
defined in Equation (31) equal zero. We repeat all experiments three times. For each repetiti
we set the length of a timecourse to $T = 30$.

3.4. Settings

For all optimization methods, we set the population size to $N = 100$. All optimiz
run maximally 1000 generations. In the case of ES, we use the initial value of σ equal
For DE, RevDE, and RevDE+, we use $F = 0.5$, and $p = 0.9$. For EDA we take $M = 1$
In the case of EDA+ and RevDE+, we use the K-NN as the surrogate model with $K =$
and we do not store more than 10,000 evaluated individuals.

4. Results & Discussion

Fitness value: In Figure 4 we present convergence of the methods in Figure 4. We not
that all methods were able to converge and achieve very similar fitness values. Hc
ever, the $(1 + 1)$-ES method was slowest due to the slow exploration capabilities. EI
also required more evaluations to obtain better results. Interestingly, DE, RevDE, RevD
and EDA+ achieved almost identical values of the fitness function (the differences we
beyond the three-digit precision). An important observation is that application of
surrogate model (the K-NN regressor) allowed to significantly speed up the conv
gence of RevDE+ and EDA+ compared to RevDE and EDA, respectively. We conclu
that all population-based methods were able to converge and achieved almost identi
scores, and our proposition of applying the surrogate model led to improving both RevI
and EDA.

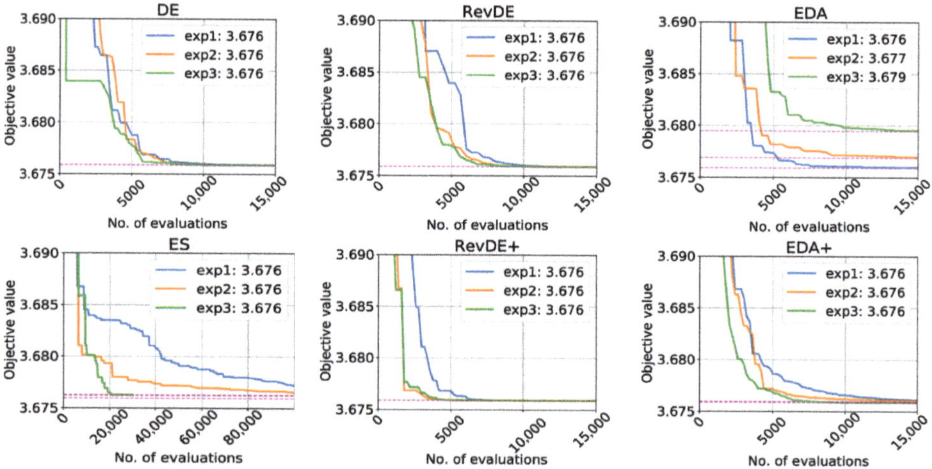

Figure 4. The convergence of the population-based optimization methods over 3 runs. In the legends, we indicate the value
of the fitness function after the methods converged.

Timecourses: The final value of the fitness function tells us how well the simu
tor models the observed timecourses for given parameters provided by an optimiz
Additionally, we can also qualitatively inspect the timecourses both the observed and
observed metabolites. In Figure 5 we present timecourses for the unobserved metaboli
for parameter values found the five methods.

For all unobserved metabolites, the average over 3 repetitions of the experiments ov
lapped with the real value or laid within the confidence interval ($3\times$ standard deviatio
This is a result that we hoped for since being able to generate unobserved metabolite
extremely important for analyzing biological systems. However, we notice that DE a

RevDE+ led to almost identical timecourses, thus, they were able to properly identify parameters.

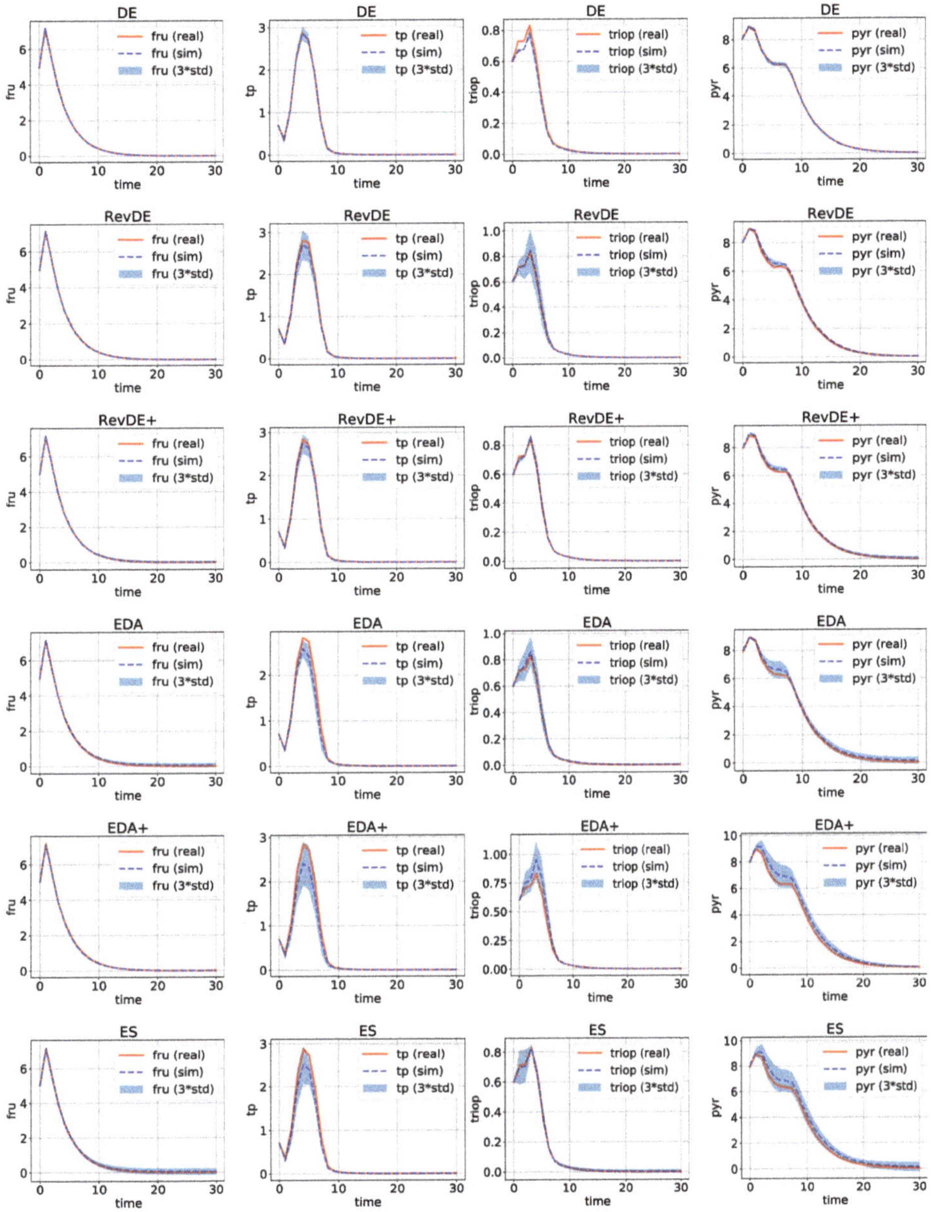

Figure 5. A comparison of the timecourses of the unobserved metabolites. Real timecourses are depicted in red, and the average value and a confidence interval ($3\times$ standard deviation) over 3 runs of the simulator is depicted in blue. The titles of the plots indicate optimization methods.

Differences in parameters: In this paper, we know precisely the values of the parameters since they were measured in [16]. Hence, we can compare the parameter values found by the optimization methods with the real parameter values. We use the absolute value of the difference of two values. We calculate the mean and the standard deviation of the difference from three runs, and use the cumulative distribution function of the folded normal distribution to visualize the distribution of differences (the ideal case is The difference between two real-valued random variable is normally distributed. However taking the absolute value of a normally distributed random variable results in the folded normal distribution.

In Figure 6 we present difference of all parameters. In general, the differences a marginal and we can conclude that all parameter values were rather properly identified. The biggest problems though appear for parameters that have very large values, e.g., k_8 k_{33}. This result is very promising because it seems to confirm the promise of the pap that it is possible to identify parameters of a complex biological network for only partial observable metabolites.

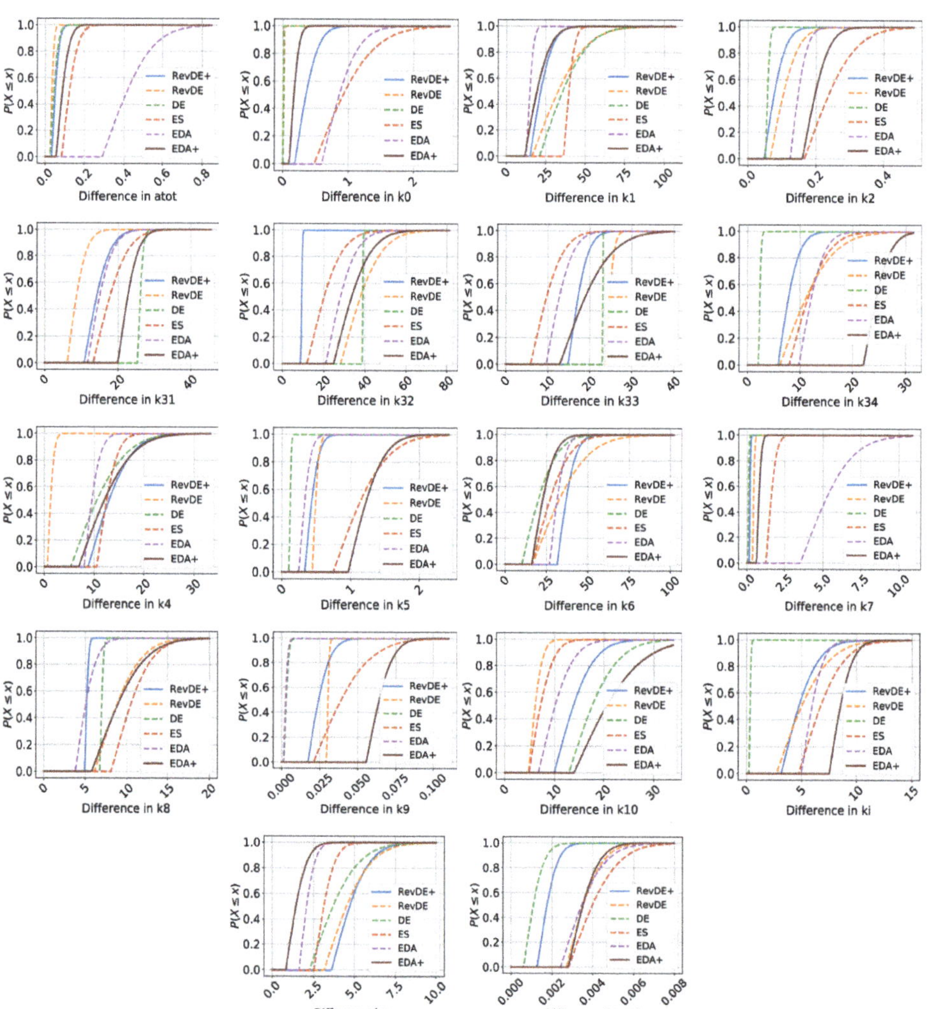

Figure 6. The cumulative distribution functions (cdfs) of the differences for all parameters. Ideally, a cdf of an optimization method should resemble a step-function centered at 0. The averages and the scales are calculated over 3 repetitions of the experiment.

5. Conclusions

In this work, we present a population-based framework for parameter identification of biological networks described as dynamic models. The obtained results indicate the great potential of population-based optimization methods in the field of biology and biochemistry. In the case of relatively low computational costs of obtaining an evaluation of parameters, the population-based methods seem to be sufficient to solve the parameter identification problem. Moreover, our results for applying surrogate models to the optimizers can be highly effective (i.e., speeding up convergence). It is a known fact (e.g., see [31,38]), nevertheless, we believe that the optimization with surrogate models has a great future and should be further investigated. For instance, considering other classes of surrogate models like Gaussian processes or (Bayesian) neural networks opens new opportunities and research questions worth following.

Additionally, the development of our framework in Python, an open-source platform, simplifies its distribution and enables its use on most operating systems. POPI4SB is easy-to-use and since the code is freely available, it constitutes a platform for developing new population-based optimizers. Therefore, the proposed framework can be relatively easily extended and serve for future research.

Author Contributions: Conceptualization, E.W.-T.; methodology, E.W.-T. and J.M.T.; software, J.M.T.; validation, E.W.-T. and J.M.T.; formal analysis, E.W.-T. and J.M.T.; investigation, E.W.-T. and J.M.T.; resources, E.W.-T. and J.M.T.; writing—original draft preparation, E.W.-T. and J.M.T.; writing—review and editing, A.E.E. and S.B.; visualization, J.M.T.; supervision, S.B.; project administration, E.W.-T. All authors have read and agreed to the published version of the manuscript.

Funding: EW-T was financed by a grant within Mobilność Plus V from the Polish Ministry of Science and Higher Education (Grant 1639/MOB/V/2017/0).

Institutional Review Board Statement: Not applicable.

Informed Consent Statement: Not applicable.

Data Availability Statement: The code is available at: https://github.com/jmtomczak/popi4sb.

Conflicts of Interest: The authors declare no conflict of interest. The funders had no role in the design of the study; in the collection, analyses, or interpretation of data; in the writing of the manuscript, or in the decision to publish the results.

References

Ingalls, B.P. *Mathematical Modeling in Systems Biology: An Introduction*; MIT Press: Cambridge, MA, USA, 2013.

Nielsen, J. Systems biology of metabolism. *Annu. Rev. Biochem.* **2017**, *86*, 245–275. [CrossRef] [PubMed]

Ideker, T.; Galitski, T.; Hood, L. A new approach to decoding life: systems biology. *Annu. Rev. Genom. Hum. Genet.* **2001**, *2*, 343–372. [CrossRef] [PubMed]

Westerhoff, H.V.; Palsson, B.O. The evolution of molecular biology into systems biology. *Nat. Biotechnol.* **2004**, *22*, 1249–1252. [CrossRef] [PubMed]

Audet, C.; Hare, W. *Derivative-Free and Blackbox Optimization*; Springer: Berlin/Heidelberg, Germany, 2017.

Larson, J.; Menickelly, M.; Wild, S.M. Derivative-free optimization methods. *arXiv* **2019**, arXiv:1904.11585.

Olivier, B.G.; Rohwer, J.M.; Hofmeyr, J.H.S. Modelling cellular systems with PySCeS. *Bioinformatics* **2005**, *21*, 560–561. [CrossRef]

Olivier, B.G.; Snoep, J.L. Web-based kinetic modelling using JWS Online. *Bioinformatics* **2004**, *20*, 2143–2144. [CrossRef]

Gatenby, R.A.; Gillies, R.J. Why do cancers have high aerobic glycolysis? *Nat. Rev. Cancer* **2004**, *4*, 891–899. [CrossRef]

Pelicano, H.; Martin, D.; Xu, R.; Huang, P. Glycolysis inhibition for anticancer treatment. *Oncogene* **2006**, *25*, 4633–4646. [CrossRef]

Duarte, N.C.; Herrgård, M.J.; Palsson, B.Ø. Reconstruction and validation of Saccharomyces cerevisiae iND750, a fully compartmentalized genome-scale metabolic model. *Genome Res.* **2004**, *14*, 1298–1309. [CrossRef]

Lee, T.I.; Rinaldi, N.J.; Robert, F.; Odom, D.T.; Bar-Joseph, Z.; Gerber, G.K.; Hannett, N.M.; Harbison, C.T.; Thompson, C.M.; Simon, I.; others. Transcriptional regulatory networks in Saccharomyces cerevisiae. *Science* **2002**, *298*, 799–804. [CrossRef]

Mensonides, F.I.; Brul, S.; Hellingwerf, K.J.; Bakker, B.M.; Teixeira de Mattos, M.J. A kinetic model of catabolic adaptation and protein reprofiling in Saccharomyces cerevisiae during temperature shifts. *Febs. J.* **2014**, *281*, 825–841. [CrossRef] [PubMed]

Nielsen, J. Yeast systems biology: model organism and cell factory. *Biotechnol. J.* **2019**, *14*, 1800421. [CrossRef] [PubMed]

Orij, R.; Urbanus, M.L.; Vizeacoumar, F.J.; Giaever, G.; Boone, C.; Nislow, C.; Brul, S.; Smits, G.J. Genome-wide analysis of intracellular pH reveals quantitative control of cell division rate by pH c in Saccharomyces cerevisiae. *Genome Biol.* **2012**, *13*, R80. [CrossRef] [PubMed]

16. Wolf, J.; Passarge, J.; Somsen, O.J.; Snoep, J.L.; Heinrich, R.; Westerhoff, H.V. Transduction of intracellular and intercellular dynamics in yeast glycolytic oscillations. *Biophys. J.* **2000**, *78*, 1145–1153. [CrossRef]

17. Jones, D.R.; Schonlau, M.; Welch, W.J. Efficient global optimization of expensive black-box functions. *J. Glob. Optim.* **1998**, *13*, 455–492. [CrossRef]

18. Eiben, A.E.; Smith, J.E. *Introduction to Evolutionary Computing*; Springer: Berlin/Heidelberg, Germany, 2015; Volume 53.

19. Gallagher, M.; Frean, M. Population-based continuous optimization, probabilistic modelling and mean shift. *Evol. Comput.* **2005**, *13*, 29–42. [CrossRef]

20. Shahriari, B.; Swersky, K.; Wang, Z.; Adams, R.P.; De Freitas, N. Taking the human out of the loop: A review of Bayesian optimization. *Proc. IEEE* **2015**, *104*, 148–175. [CrossRef]

21. Bäck, T.; Foussette, C.; Krause, P. *Contemporary Evolution Strategies*; Springer: Berlin/Heidelberg, Germany, 2013.

22. Moré, J.J.; Wild, S.M. Benchmarking derivative-free optimization algorithms. *SIAM J. Optim.* **2009**, *20*, 172–191. [CrossRef]

23. Schwefel, H.P. *Numerische Optimierung von Computer-Modellen Mittels der Evolutionsstrategie*; Springer: Berlin/Heidelberg, Germany, 1977.

24. Storn, R.; Price, K. Differential evolution—A simple and efficient heuristic for global optimization over continuous spaces. *J. Glob. Optim.* **1997**, *11*, 341–359. [CrossRef]

25. Price, K.; Storn, R.M.; Lampinen, J.A. *Differential Evolution: A Practical Approach to Global Optimization*; Springer Science & Business Media: Berlin/Heidelberg, Germany, 2006.

26. Pedersen, M.E.H. *Good Parameters for Differential Evolution*; Technical Report HL1002; Hvass Laboratories. 2010. Available online: https://citeseerx.ist.psu.edu/viewdoc/download?doi=10.1.1.298.2174&rep=rep1&type=pdf (accessed on 4 January 2021).

27. Tomczak, J.M.; Weglarz-Tomczak, E.; Eiben, A.E. Differential Evolution with Reversible Linear Transformations. *arXiv* **2020**, arXiv:2002.02869.

28. Larrañaga, P.; Lozano, J.A. *Estimation of Distribution Algorithms: A New Tool for Evolutionary Computation*; Springer Science & Business Media: Berlin/Heidelberg, Germany, 2001.

29. Mühlenbein, H.; Paass, G. From recombination of genes to the estimation of distributions I. Binary parameters. In *International Conference on Parallel Problem Solving from Nature*; Springer: Berlin/Heidelberg, Germany, 1996; pp. 178–187.

30. Pelikan, M.; Hauschild, M.W.; Lobo, F.G. Estimation of distribution algorithms. In *Springer Handbook of Computational Intelligence*; Springer: Berlin/Heidelberg, Germany, 2015; pp. 899–928.

31. Jin, Y. Surrogate-assisted evolutionary computation: Recent advances and future challenges. *Swarm Evol. Comput.* **2011**, *1*, 61–70. [CrossRef]

32. Hynne, F.; Danø, S.; Sørensen, P.G. Full-scale model of glycolysis in Saccharomyces cerevisiae. *Biophys. Chem.* **2001**, *94*, 121–163. [CrossRef]

33. Kourdis, P.D.; Goussis, D.A. Glycolysis in saccharomyces cerevisiae: algorithmic exploration of robustness and origin of oscillations. *Math. Biosci.* **2013**, *243*, 190–214. [CrossRef]

34. Teusink, B.; Passarge, J.; Reijenga, C.A.; Esgalhado, E.; Van der Weijden, C.C.; Schepper, M.; Walsh, M.C.; Bakker, B.M.; Van Dam, K.; Westerhoff, H.V.; et al. Can yeast glycolysis be understood in terms of in vitro kinetics of the constituent enzymes? Testing biochemistry. *Eur. J. Biochem.* **2000**, *267*, 5313–5329. [CrossRef]

35. Van Eunen, K.; Bouwman, J.; Daran-Lapujade, P.; Postmus, J.; Canelas, A.B.; Mensonides, F.I.; Orij, R.; Tuzun, I.; Van Den Brink, J.; Smits, G.J.; et al. Measuring enzyme activities under standardized in vivo-like conditions for systems biology. *FEBS J.* **2010**, *277*, 749–760. [CrossRef] [PubMed]

36. Available online: https://jjj.bio.vu.nl/models/wolf/ (accessed on 7 August 2020).

37. Cranmer, K.; Brehmer, J.; Louppe, G. The frontier of simulation-based inference. *Proc. Natl. Acad. Sci. USA* **2020**, *117*, 30055–30062. [CrossRef] [PubMed]

38. Gatopoulos, I.; Lepert, R.; Wiggers, A.; Mariani, G.; Tomczak, J. Evolutionary Algorithm with Non-parametric Surrogate Model for Tensor Program optimization. In Proceedings of the 2020 IEEE Congress on Evolutionary Computation (CEC), Glasgow, UK, 19–24 July 2020.

 *processes*

Article

A Hybrid of Particle Swarm Optimization and Harmony Search to Estimate Kinetic Parameters in *Arabidopsis thaliana*

Mohamad Saufie Rosle [1], Mohd Saberi Mohamad [2,3,*], Yee Wen Choon [2,3], Zuwairie Ibrahim [4], Alfonso González-Briones [5,6], Pablo Chamoso [5] and Juan Manuel Corchado [5]

[1]   Faculty of Computing, Universiti Teknologi Malaysia, Skudai 81310, Malaysia; msrsufi@gmail.com
[2]   Institute for Artificial Intelligence and Big Data, Universiti Malaysia Kelantan, Kota Bharu 16100, Malaysia; ewenchoon@gmail.com
[3]   Faculty of Bioengineering and Technology, Universiti Malaysia Kelantan, Jeli 17600, Malaysia
[4]   College of Engineering, Universiti Malaysia Pahang, Lebuhraya Tun Razak, Gambang, Kuantan 26300, Malaysia; zuwairie@ump.edu.my
[5]   BISITE Research Group, IBSAL, Digital Innovation Hub, University of Salamanca, Edificio I+D+i, C/Espejos s/n, 37007 Salamanca, Spain; alfonsogb@usal.es (A.G.-B.); chamoso@usal.es (P.C.); corchado@usal.es (J.M.C.)
[6]   Research Group on Agent-Based, Social and Interdisciplinary Applications (GRASIA), Complutense University of Madrid, 28040 Madrid, Spain
*   Correspondence: saberi@umk.edu.my

Received: 30 April 2020; Accepted: 20 July 2020; Published: 2 August 2020

Abstract: Recently, modelling and simulation have been used and applied to understand biological systems better. Therefore, the development of precise computational models of a biological system is essential. This model is a mathematical expression derived from a series of parameters of the system. The measurement of parameter values through experimentation is often expensive and time-consuming. However, if a simulation is used, the manipulation of computational parameters is easy, and thus the behaviour of a biological system model can be altered for a better understanding. The complexity and nonlinearity of a biological system make parameter estimation the most challenging task in modelling. Therefore, this paper proposes a hybrid of Particle Swarm Optimization (PSO) and Harmony Search (HS), also known as PSOHS, designated to determine the kinetic parameter values of essential amino acids, mainly aspartate metabolism, in *Arabidopsis thaliana*. Three performance measurements are used in this paper to evaluate the proposed PSOHS: the standard deviation, nonlinear least squared error, and computational time. The proposed algorithm outperformed the other two methods, namely Simulated Annealing and the downhill simplex method, and proved that PSOHS is a more suitable algorithm for estimating kinetic parameter values.

Keywords: Particle Swarm Optimization; Harmony Search; parameter estimation; *Arabidopsis thaliana*

---

1. Introduction

In silico optimization is a rapid and cost-effective method for finding optimal solutions in optimization problems. The two currently available methods are local and global methods. Each, however, comes with inherent systematic problems that require troubleshooting. The local optimization method performs poorly when solving non-linear and dynamic biological problems [1]. In this type of problem, the global search is a more effective method as it is capable of finding an optimal solution that satisfies all requirements. In fact, the global search method has recently been subject of much attention within the scientific community [1–3]. Parameter estimation is an essential phase in the simulation of the biological system because the value of the parameter determines the behaviour of the model. In estimating parameter values, it is important to first determine the objective function that can then be minimized by using suitable optimization methods. The objective function must be minimized to obtain the ideal value of the parameter. Several optimization methods have, therefore, been applied to estimate parameter values [4]. After comparing several methods, Baker et al. [4] pointed out that while the Genetic Algorithm (GA) was less time-consuming than Simulated Annealing (SA), it faced the local minima problem. In MCMC approaches are often applied to determine the posterior distribution of rate parameters. However, developing a good MCMC sampler for its multimodal and dimensional parameter distribution is challenging. Valderrama-Bahamóndez and Fröhlich [5] found that parallel adaptive MCMC performed better in parameter estimations after comparing the ability of different MCMC approaches to estimate the kinetic rate parameters of ordinary differential equation (ODE) systems.

In addition, global optimization algorithms, like Particle Swarm Optimization (PSO) [6,7], SA [8], and Scatter Search [9,10], have been successful in the estimation of parameters in different biological models. In a comparative study on computational intelligence methods, carried out by Tangherloni et al. [11], the performance of various meta-heuristics was compared; in particular, their ability to estimate the parameters with a set of benchmark functions and synthetic biochemical models. Tangherloni et al. [11] concluded that classic benchmark functions lacked the comprehensibility that compounded the real-world optimization problem. The hybridization of optimization methods has been proposed so that the best features of each method could be utilized. Convergence to global optima is a consistent problem in standard PSO for multimodal and high-dimensional functions [12]. Hence, Fu et al. [12] proposed a hybrid method by incorporating the evolutionary operations of the Differential Evolution (DE) algorithm to improve its conventional velocity updating strategy in PSO. Harmony Search (HS) is a simple method with an efficient and evolutionary algorithm. It has parameters like the Harmony Memory Considering Rate (HMCR) and Pitch Adjusting Rate (PAR) that solve the local optima problem. Furthermore, HS is able to balance between intensification and diversification whereby promising regions and non-explored regions are thoroughly and evenly explored in both processes [13]. This paper introduces a hybrid of Particle Swarm Optimization [14] and Harmony Search [15] (PSOHS) and simulates the essential amino acid metabolism for kinetic parameter estimation.

PSO estimates kinetic parameter values by using swarm intelligence. The candidate solutions are analogous to particles flying with specific velocities in specific directions in the search space, either alone or together with their companions. Presumably, the particle will move to its best positions. HS, on the other hand, is based on the analogy with natural musical performance processes. By improving the pitches of the instrument of each music player, a better harmony is achieved and, hence, the quality of the performance [16]. Gao et al. [16] further discussed this method and its applications. The historical development of the HS algorithm structure had been extensively reviewed by Zhang and Geem [17].

This paper introduces a new hybrid algorithm, PSOHS, into the SBToolbox to estimate kinetic parameters by simulating the aspartate metabolism of isoleucine, lysine, and threonine in a small plant of the mustard family, *Arabidopsis thaliana*. Its model is often chosen in research because it has been demonstrated to render better results when analysing plant development and growth. This paper focuses on the biochemical reactions of essential amino acid production in *Arabidopsis thaliana*.

2. Materials and Methods

This section describes and discusses the problem formulation, and it details the hybridization of PSO and HS. A basic PSO is also described to differentiate between the basic PSO and the hybrid version.

2.1. Problem Formulation

This paper uses a biochemical system, which is a mathematical modelling framework based on ordinary differential equation (ODE). In the system, biochemical processes are represented using power-law expansions in the variables. In a biochemical process, a system of kinetic equations is formed according to the information regarding the underlying network structure of a pathway, and the parameters are acquired from the literature or estimated values are acquired from a data fit. Chemical kinetics mean rates of chemical reactions. The parameter estimation problem is posed in Equation (1) below. There, $s(X)$ denotes a biological compound, depending on a set of parameters $X = (X1,X2,X3,\ldots Xd)$, where d is the total number of parameters. Hence, the reaction rate of the compound s is presented by

$$\begin{aligned}
\frac{ds}{dt} &= g(s(X),t),\\
s(t_0) &= s(0),\\
y &= g(s(X),t) + e.
\end{aligned} \tag{1}$$

In Equation (1), g represents a nonlinear function and t represents the sampling time. Then, y is a time series of simulated data, also known as the output of the model, and e represents the noise data that is randomly produced by Gaussian noise n (1,0). The purpose of the parameter estimation is to discover the set of the optimal parameter, which is denoted as X. Then the variance between the simulated time-series data denoted as y and experimental time-series data denoted as y^{exp} can be reduced. The variance is calculated by applying the nonlinear least squared error function, $f(X)$, which is shown in Equation (2):

$$f(X) = min \sum_{i=1}^{n} (y^{exp} - y)^2 \tag{2}$$

In Equation (2), n is the total number that maximum value generated and i is the index variable.

2.2. Particle Swarm Optimization (PSO)

PSO is a swarm intelligence-based optimization algorithm. The method was developed by Kennedy and Eberhart [6] and had been inspired by the social behaviour of flocking birds and schooling fish when searching for food [6].

In PSO, particles act as a possible solution in the search space of a problem. Every particle in the search space is assigned a certain velocity so that its movements through the search space can be determined. In the search space, each particle's movement is affected not only by its local best-known position, but it is also directed toward the best-known positions, where the best conditions are those that have been found by other particles.

Hence, there are two positions in the whole swarm, the local best-known position and the best-known position in the swarm. *pbest* represents the local best-known position while *gbest* represents the best-known positions in the swarm.

Figure 1 shows the PSO flowchart. First, each particle is assigned a random position within the problem space to form an initial population. Then, the fitness of each particle is evaluated and compared with *pbest*. If the current value is better than *pbest*, then this value is updated to the new *pbest*. Then, the particle's best-known position and the swarm's best-known position are updated. The termination criterion is when the maximum number of iterations is performed, or a solution meets the adequate objective function value.

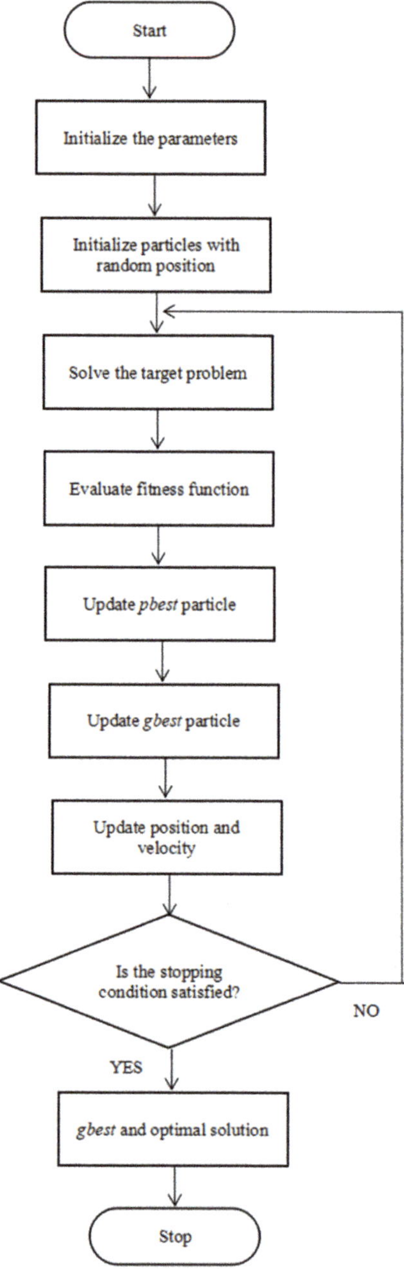

Figure 1. Flowchart of Particle Swarm Optimization (PSO).

2.3. Harmony Search (HS)

Geem et al. [13] developed HS after being inspired by the music improvisation process. The algorithm achieved better harmony the same way as music players improved the pitches of their instruments [18]. To escape from the local minima, HS performs a stochastic random search instead of a gradient search. There are three rules to generate perfect harmony, which are Random Selection (RS), Harmony Memory Considering Rate (HMCR), and Pitching Adjust Rate (PAR). Figure 2 shows the flowchart of HS. First, a population of random harmonies in Harmony Memory (HM) is initialized. Several randomly generated solutions are included in the initial HM. The next step is to improvise a new solution from the HM followed by updating the latter. In each repetition, the algorithm improvises a new harmony, and the latest harmony is generated by the following three rules: (1) HMCR is used to select the variables of the new harmonies from the overall HM harmonies; (2) PA is responsible for local improvement; and (3) RS provides random elements for the new harmony. If the latest harmony is better than the existing one, then it evaluates and replaces the worst harmony in HM. This process is iterated until the stopping criteria are met.

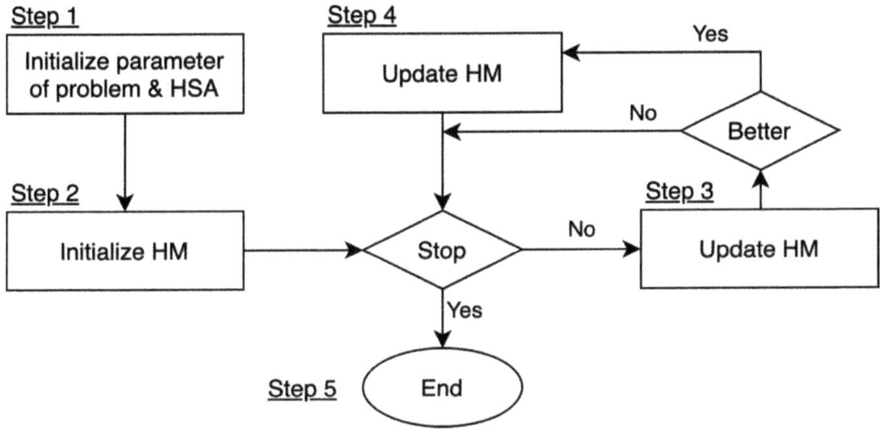

Figure 2. Flowchart of Harmony Search (HS).

2.4. A Hybrid of PSO and HS (PSOHS)

PSOHS is proposed in this paper, which is a hybrid of HS and PSO. Figure 3 shows the steps taken by the proposed algorithm in order to acquire the optimal parameter values.

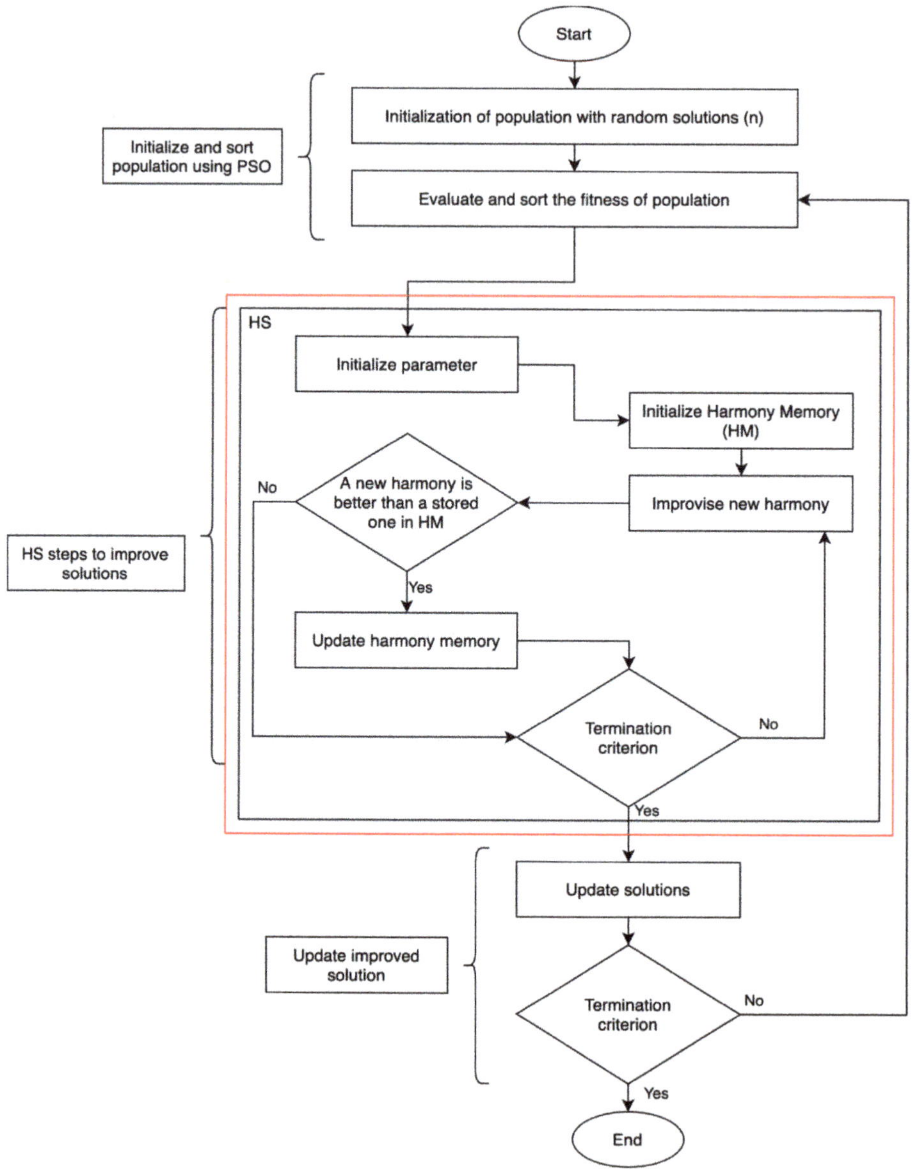

Figure 3. Steps in PSOHS for estimating parameter value. The red box is Harmony Search, which is hybridized with PSO to improve its performance.

2.4.1. Initialization

The first step of PSOHS is the initialization of the population of random solutions with random velocities and positions in d-dimensions in the search space. Some parameter values of the proposed algorithm are assigned: (i) harmony memory size; (ii) harmony consideration rate; (iii) swarm size; (iv) positive constant; (v) the number of iterations; and (vi) inertia weight. The harmony memory size is

6, the number of iterations is 20, and the harmony consideration rate is 0.9. Then, the fitness of each particle in d variables is evaluated.

2.4.2. Iteration

The next step involves evaluating and sorting the fitness of the population. In each iteration, two values are updated, which are *pbest* and *gbest*. At the iteration step, each solution is evaluated by using the optimization fitness function. If the current value of the solutions in the d-dimensional space is better than *pbest*, then the pbest value is updated to the current value and the *pbest* location is the current location. Then, the fitness value is compared with the best-known positions in the swarm. If the current value is better than *gbest*, then *gbest* is updated to the current solution index.

2.4.3. Hybridization of Harmony Search

First, the parameters and the HS Memory (HM) are initialized. The parameters are (i) the size of HM; (ii) the Harmony Memory Considering Rate (HMCR); and (iii) the Pitching Adjust Rate (PAR). Next, the latest harmony is created based on three rules: (i) Random Selection (RS); (ii) HMCR; and (iii) PAR. Firstly, several randomly created solutions to the problems are included in the initial HM. Each component of this solution is acquired on the basis of the HMCR. The probability of choosing a component from the HM members is defined as HMCR and, therefore, 1-HMCR is the probability of creating it randomly. The harmony is chosen from a random HM member and PAR is used to further mutate the chosen harmony. The probability of a candidate from the HM to be mutated is determined by PAR. The RS is responsible for providing random elements to the new harmony. If the latest harmony yields a better fitness, it will replace the worst member in the HM. Otherwise, it is removed. This process is iterated until a stopping condition is reached.

2.4.4. Termination

The iterations terminate if the stopping criterion is achieved. The two stopping criteria are when the fitness function can no longer improve or when the predefined maximum loop values are reached.

3. Experimental Setup

The experiments have been carried out by means of computer simulation in which the performance of the selected algorithms has been assessed.

Experiment Setup (Computational Approach)

Three different algorithms; PSOHS, SA, and the downhill simplex method, have been used to estimate the parameter values. The algorithms were executed in MATLAB R2010a on a 1024 MB (1 GB) RAM and Intel Pentium 4 processor laptop. The result of PSOHS was compared with SA and the downhill simplex method. The total run for estimating all the kinetic values was 50 individual runs. The accuracy, consistency, nonlinear least squared error, and standard deviation were calculated and compared with both algorithms to evaluate the performance of PSOHS. The formula for calculating the nonlinear least squared error and standard deviation is given below:

$$e = \sum_{i=1}^{N} (y - y_i)^2 \tag{3}$$

In Equation (3), e means the squares of the errors, y is the measurement result, and y_i is the simulated result. Equation (4) is used to calculate the average squared error, where A represents the average squared error and N represents the number of samples.

$$A = \frac{e}{N} \tag{4}$$

Equation (5) is used to calculate the standard deviation.

$$STD = \sqrt{\frac{e}{N}}$$

(5)

Aspartate metabolism from *Arabidopsis thaliana* is the dataset that has been used in this paper. *Arabidopsis thaliana* has been chosen as a model organism due to its advantages for genetic experiments, such as (i) a short generation time; (ii) its small size; and (iii) its prolific seed production. In microorganisms, aspartate acts as the precursor to several amino acids, including methionine, threonine, isoleucine, and lysine, which are essential for humans. Threonine, isoleucine, and lysine have been selected in this paper because all of them cannot be produced by the body, but they are important in almost all body functions. Dataset details are shown in Table 1. In this paper, the values of a total of 31 kinetic parameters are estimated using PSOHS.

Table 1. Information on the dataset.

Dataset	Aspartate Metabolism
Plant model	*Arabidopsis thaliana*
Download link	https://www.ebi.ac.uk/biomodels-main/BIOMD0000000212 [19]

4. Result and Discussion

The performance of PSOHS was compared with SA and the downhill simplex method. Tables 2–4 show the kinetic parameter values that are estimated using PSOHS, SA, and the downhill simplex method on the basis of the experimental value [19]. The parameter values might range wildly in scale as they originate from the previous work [19]. It should be noted that Equation (3) is used to calculate the distance between the experimental data and model simulation for each reaction in each amino acid (isoleucine, lysine, and threonine). There are many reactions involved in each amino acid as well as ODEs in *Arabidopsis thaliana*. Tables 2–4 summarize the experimental results. The average squared error in Tables 2–4 shows the average reaction in the ODE system for each amino acid that involves a number of kinetic parameters. Meanwhile, Tables 5–7 report the parameter values obtained from the experimental data, as well as those generated from the proposed PSOHS, the downhill simplex method, and SA.

Table 2. Comparison between PSOHS, the downhill simplex method, and Simulated Annealing (SA) in estimating six parameters for isoleucine in terms of computational time, average squared error, and standard deviation.

Algorithms Measurements	PSOHS	Downhill Simplex Method	SA
Computational time (seconds)	**100.23**	130.56	778.00
Average squared error, A	**0.0003**	0.0008	0.0012
Standard deviation, STD	**0.0002**	0.0004	0.002

Note: The bold numbers represent the best result.

Table 3. Comparison between the performance of PSOHS, downhill simplex method, and SA in estimating nine parameters for lysine in terms of computational time, average squared error, and standard deviation.

Algorithms Measurements	PSOHS	Downhill Simplex Method	SA
Computational time (seconds)	**184.03**	376.59	1518.05
Average squared error, A	**0.0211**	0.084	0.0406
Standard deviation, STD	**0.0133**	0.0998	0.0347

Note: The best results are in bold.

Table 4. Comparison between the performance of PSOHS, the downhill simplex method, and SA in estimating sixteen parameters for threonine in terms of computational time, average squared error, and standard deviation.

Algorithms Measurements	PSOHS	Downhill Simplex Method	SA
Computational time (seconds)	**255.37**	362.32	1794.91
Average squared error, A	**0.0024**	0.012	0.0066
Standard deviation, STD	**0.0037**	0.017	0.0079

Note: The best results are in bold.

Table 5. List of kinetic parameter values for isoleucine with the experimental values.

Parameters	Experimental [19]	PSOHS	Downhill Simplex Method	SA
Vtd_TD_k_app_exp	0.0124	**0.0101**	0.0138	0.0156
Vtd_TD_Ile_Ki_no_Val_app_exp	30	55.62	**33.15**	59.995
Vtd_TD_Val_Ka1_app_exp	73	154.02	**74.18**	196.46
Vtd_TD_Val_Ka2_app_exp	615	812.01	**686.34**	3014.68
Vtd_TD_nH_app_exp	3	**5.27**	6.30	15.28
VileTRNA_Ile_tRNAS_Ile_Km	20	**29.55**	32.25	31.86

Note: The best results are in bold.

Table 6. List of kinetic parameter values for lysine with the experimental values.

Parameters	Experimental [19]	PSOHS	Downhill Simplex Method	SA
Vdhdps1_DHDPS1_k_app_exp	1	**1.19**	1.26	1.24
Vdhdps1_DHDPS1_Lys_Ki_app_exp	10	**10.76**	11.47	12.67
Vdhdps1_DHDPS1_nH_exp	2	**2.02**	2.71	2.82
Vdhdps2_DHDPS2_k_app_exp	1	**1.001**	0.95	0.9
Vdhdps2_DHDPS2_Lys_Ki_app_exp	33	**32.23**	34.96	35.48
Vdhdps2_DHDPS2_nH_exp	2	**2.004**	3.68	2.38
VlysTRNA_Lys_tRNAS_Lys_Km	25	**26.47**	31.39	27.19
VlysKR_LKR_kcat_exp	3.1	**3.06**	3.04	3.71
VlysKR_LKR_Lys_Km_exp	13,000	**13,000.11**	9681.89	14,258.63

Note: The bold numbers represent the best result.

Table 7. List of kinetic parameter values for threonine with the experimental values.

Parameters	Experimental [19]	PSOHS	Downhill Simplex Method	SA
Vts1_TS1_kcatmin_exp	0.42	**0.401**	0.78	0.86
Vts1_TS1_AdoMet_kcatmax_exp	3.5	**3.82**	5.91	6.29
Vts1_TS1_nH_exp	2	**2.00**	1.98	1.93
Vts1_TS1_AdoMet_Ka1_exp	73	**71.23**	85.86	181.73
Vts1_TS1_AdoMEt_Km_no_AdoMet_exp	250	**250.04**	236.43	551.08
Vts1_TS1_AdoMet_Ka2_exp	0.5	**0.50**	0.55	1.23
Vts1_TS1_AdoMet_Ka3_exp	1.09	**1.095**	1.72	2.22
Vts1_TS1_AdoMet_Ka4_exp	140	**150.42**	172.29	336.05
Vts1_TS1_Phosphate_Ki_exp	1000	**1001.46**	1066.50	2823.02
Vtd_TD_k_app_exp	0.0124	**0.0124**	0.0126	0.0166
Vtd_TD_Ile_Ki_no_Val_app_exp	30	**31.12**	89.33	21.72
Vtd_TD_Val_Ka1_app_exp	73	**73.55**	99.32	59.69
Vtd_TD_Val_Ka2_app_exp	615	**617.62**	1202.39	2949.7
Vtd_TD_nH_app_exp	3	**3.11**	7.41	6.93
Vtha_THA_kcat_exp	1.7	**1.71**	4.74	4.37
Vtha_THA_Thr_Km_exp	7100	**7100.63**	12,238.29	18,663.59

Note: The bold numbers represent the best result.

This paper focuses on parameter estimation using the proposed PSOHS. The performance of PSOHS was measured in terms of computational time, model accuracy, and precision of the algorithms. Model accuracy is measured by the distance value between the experimental data and model simulation using the nonlinear least squared method. In addition, the average squared error was calculated to get the average of all ODEs in each amino acid. To test the algorithm's precision, standard deviation was used for 50 individual runs. High standard deviations demonstrate low precision, and low standard deviations demonstrate high precision. The experiments were carried out in 50 individual runs to test the algorithms, and the result shown is the best multivariate solution among the runs. The average squared error and standard deviation were calculated from the runs. Tables 2–4 show the comparisons of computational time in seconds, average squared error, and standard deviation among PSOHS, SA, and the downhill simplex method. Table 2 shows the execution time for PSOHS to estimate the six kinetic parameters of isoleucine is 100.23 s, which is the lowest compared to 130.56 s for the downhill simplex method and 778.00 s for SA. The standard deviation of PSOHS is 0.0002, which is the closest to zero compared to the downhill simplex method and SA, which are 0.0004 and 0.002. Based on these comparisons, PSOHS shows the lowest average squared error and a low standard deviation, and this proves that PSOHS is more consistent, precise, and reliable in parameter values estimation compared to SA and the downhill simplex method.

Table 3 shows that the average nonlinear least squared error for the three algorithms is 0.0211, 0.084, and 0.0406. The standard deviations of the three algorithms are 0.0133, 0.0998, and 0.0347. Besides, the computational time for PSOHS is 184.03; for the downhill simplex method it is 376.59 and for SA it is 1518.05. From among the three algorithms for estimating the kinetic parameter of lysine, PSOHS shows the best result.

Threonine has the greatest number of kinetic parameters to be estimated among the selected amino acids. Table 4 presents the comparison among PSOHS, the downhill simplex method, and SA. It seems that the average nonlinear least squared error for PSOHS is smaller than the downhill simplex method and SA. The average nonlinear least squared error for PSOHS is 0.0024, while the average nonlinear least squared errors for the downhill simplex method and SA are 0.012 and 0.0066, respectively.

Furthermore, to estimate all kinetic parameters, PSOHS takes 255.37 s, which is a considerably lower computational time than the downhill simplex method, which takes 362.32 s, and SA, which takes 1794.91 s. In terms of standard deviation, the downhill simplex method and SA show a high standard deviation compared to PSOHS. The standard deviation for PSOHS, downhill simplex method, and SA are 0.0037, 0.017, and 0.0079, respectively. The results show that PSOHS outperforms the downhill simplex method and SA in estimating the sixteen kinetic parameters of threonine.

5. Conclusions

In conclusion, when estimating kinetic parameter values, PSOHS performed better than the downhill simplex method and SA, as shown by the smaller standard deviation in PSOHS. Moreover, PSOHS is less time-consuming. The lower nonlinear least squared error of PSOHS also proves that this algorithm is more accurate compared to SA and the simplex downhill method. Future lines of research are going to focus on including different performance measurements and algorithms and on comparing how they affect the performance of PSOHS. Only one dataset has been used in this research due to an unavoidable constraint. However, more datasets can be used in future work. Large-scale metabolic parameter estimation is preferable. However, the inclusion of more datasets poses a bigger challenge in that the parameters of every single gene and its product will be needed in order to be estimated. It leads to large-scale metabolic parameter estimations [20]. PSOHS should be further expanded to that scale with the aim of resolving the problem. Besides, the shortcomings of the existing HS, such as the fine-tuning ability of the algorithm, can also be improved in future work [21].

Author Contributions: Conceptualization, M.S.R. and M.S.M.; methodology, M.S.R. and M.S.M.; project administration, M.S.M.; resources, M.S.R.; software, M.S.R.; supervision, M.S.M.; writing—original draft, M.S.R.; writing—review and editing, M.S.M., Y.W.C., Z.I., A.G.-B., P.C. and J.M.C. All authors have read and agreed to the published version of the manuscript.

Funding: We would like to thank the Skim Geran Penyelidikan Fundamental (FRGS-MRSA) (no grant: R/FRGS/A0800/01655A/003/2020/00720) from Ministry of Education Malaysia for their support in order to make this research a success.

Conflicts of Interest: The authors declare no conflicts of interest.

References

1. Parsopoulos, K.E.; Vrahatis, M.N. *Particle Swarm Optimization and Intelligence: Advances and Applications*; IGI Global: Hershey, PA, USA, 2010.
2. Sun, J.; Garibaldi, J.M.; Hodgman, C. Parameter Estimation Using Metaheuristics in Systems Biology: A Comprehensive Review. *IEEE/ACM Trans. Comput. Biol. Bioinform.* **2012**, *9*, 185–202. [CrossRef] [PubMed]
3. Horst, R.; Pardalos, P.M.; Thoai, N.V. *Introduction to Global Optimization*; Kluwer Academic Publishers: Dordrecht, The Netherlands, 2013.
4. Baker, S.M.; Schallau, K.; Junker, B.H. Comparison of different algorithms for simultaneous estimation of multiple parameters in kinetic metabolic models. *J. Integr. Bioinform.* **2010**, *7*, 254–262. [CrossRef]
5. Valderrama-Bahamóndez, G.I.; Fröhlich, H. MCMC Techniques for Parameter Estimation of ODE Based Models in Systems Biology. *Front. Appl. Math. Stat.* **2019**, *5*, 55. [CrossRef]
6. Kennedy, J.; Eberhart, R. Particle swarm optimization. In Proceedings of the ICNN'95—International Conference on Neural Networks, Perth, WA, Australia, 27 November–1 December 1995; Volume 4, pp. 1942–1948.
7. Campbell, K.S. Interactions between Connected Half-Sarcomeres Produce Emergent Mechanical Behavior in a Mathematical Model of Muscle. *PLoS Comput. Biol.* **2009**, *5*, e1000560. [CrossRef] [PubMed]
8. Kirkpatrick, S.; Gelatt, C.D.; Vecchi, M.P. Optimization by simulated annealing. *Science* **1983**, *220*, 671–680. [CrossRef] [PubMed]
9. Remli, M.A.; Deris, S.; Mohamad, M.S.; Omatu, S.; Corchado, J.M. An enhanced scatter search with combined opposition-based learning for parameter estimation in large-scale kinetic models of biochemical systems. *Eng. Appl. Artif. Intell.* **2017**, *62*, 164–180. [CrossRef]

10. Remli, M.A.; Mohamad, M.S.; Deris, S.; Samah, A.A.; Omatu, S.; Corchado, J.M. Cooperative enhanced scatter search with opposition-based learning schemes for parameter estimation in high dimensional kinetic models of biological systems. *Expert Syst. Appl.* **2019**, *116*, 131–146. [CrossRef]

11. Tangherloni, A.; Spolaor, S.; Cazzaniga, P.; Besozzi, D.; Rundo, L.; Mauri, G.; Nobile, M.S. Biochemical parameter estimation vs. benchmark functions: A comparative study of optimization performance and representation design. *Appl. Soft Comput.* **2019**, *81*, 105494. [CrossRef]

12. Fu, W.; Johnston, M.; Zhang, M. Hybrid Particle Swarm Optimisation Algorithms Based on Differential Evolution and Local Search. In Proceedings of the AI 2010: Advances in Artificial Intelligence Lecture Notes in Computer Science, Adelaide, Australia, 7–10 December 2010; pp. 313–322. [CrossRef]

13. Geem, Z.W.; Kim, J.H.; Loganathan, G. A New Heuristic Optimization Algorithm: Harmony Search. *Simulation* **2001**, *76*, 60–68. [CrossRef]

14. Ng, S.T.; Chong, C.K.; Choon, Y.W.; Chai, L.E.; Deris, S.; Illias, R.M.; Shamsir, M.S.; Mohamad, M.S. Estimating Kinetic Parameters for Essential Amino Acid Production in Arabidopsis Thaliana by Using Particle Swarm Optimization. *J. Teknol.* **2013**, *64*. [CrossRef]

15. Geem, Z.W. *Recent Advances in Harmony Search Algorithm*; Springer Science & Business Media: Berlin/Heidelberg, Germany, 2010; Volume 270.

16. Gao, X.Z.; Govindasamy, V.; Xu, H.; Wang, X.; Zenger, K. Harmony search method: Theory and applications. *Comput. Intell. Neurosci.* **2015**. [CrossRef] [PubMed]

17. Zhang, T.; Geem, Z.W. Review of harmony search with respect to algorithm structure. *Swarm Evol. Comput.* **2019**, *48*, 31–43. [CrossRef]

18. Lee, K.S.; Geem, Z.W.; Lee, S.H.; Bae, K.W. The harmony search heuristic algorithm for discrete structural optimization. *Eng. Optim.* **2005**, *37*, 663–684. [CrossRef]

19. Curien, G.; Bastien, O.; Robert-Genthon, M.; Cornish-Bowden, A.; Cárdenas, M.L.; Dumas, R. Understanding the regulation of aspartate metabolism using a model based on measured kinetic parameters. *Mol. Syst. Biol.* **2009**, *5*, 271. [CrossRef] [PubMed]

20. Mason, J.C.; Covert, M.W. An energetic reformulation of kinetic rate laws enables scalable parameter estimation for biochemical networks. *J. Theor. Biol.* **2019**, *461*, 145–156. [CrossRef] [PubMed]

21. Zhu, Q.; Tang, X.; Li, Y.; Yeboah, M.O. An improved differential-based harmony search algorithm with linear dynamic domain. *Knowl. -Based Syst.* **2020**, *187*, 104809. [CrossRef]

 *processes*

Article

MPPIF-Net: Identification of Plasmodium Falciparum Parasite Mitochondrial Proteins Using Deep Features with Multilayer Bi-directional LSTM

Samee Ullah Khan [1] and Ran Baik [2,*]

[1] Intelligent Media Laboratory, Digital Contents Research Institute, Sejong University, Seoul 143-747, Korea; sameek3797@gmail.com

[2] Department of Computer Engineering, Convergence School of ICT, Honam University, #417 Eodeung-daero, Gwangsan-gu, Gwangju 506-090, Korea

* Correspondence: baik@honam.ac.kr or ranbaik@gmail.com

Received: 29 April 2020; Accepted: 15 June 2020; Published: 22 June 2020

Abstract: Mitochondrial proteins of Plasmodium falciparum (MPPF) are an important target for anti-malarial drugs, but their identification through manual experimentation is costly, and in turn, their related drugs production by pharmaceutical institutions involves a prolonged time duration. Therefore, it is highly desirable for pharmaceutical companies to develop computationally automated and reliable approach to identify proteins precisely, resulting in appropriate drug production in a timely manner. In this direction, several computationally intelligent techniques are developed to extract local features from biological sequences using machine learning methods followed by various classifiers to discriminate the nature of proteins. Unfortunately, these techniques demonstrate poor performance while capturing contextual features from sequence patterns, yielding non-representative classifiers. In this paper, we proposed a sequence-based framework to extract deep and representative features that are trust-worthy for Plasmodium mitochondrial proteins identification. The backbone of the proposed framework is MPPF identification-net (MPPFI-Net), that is based on a convolutional neural network (CNN) with multilayer bi-directional long short-term memory (MBD-LSTM). MPPIF-Net inputs protein sequences, passes through various convolution and pooling layers to optimally extract learned features. We pass these features into our sequence learning mechanism, MBD-LSTM, that is particularly trained to classify them into their relevant classes. Our proposed model is experimentally evaluated on newly prepared dataset PF2095 and two existing benchmark datasets i.e., PF175 and MPD using the holdout method. The proposed method achieved 97.6%, 97.1%, and 99.5% testing accuracy on PF2095, PF175, and MPD datasets, respectively, which outperformed the state-of-the-art approaches.

Keywords: mitochondrial protein; machine learning; bi-directional LSTM; plasmodium falciparum

1. Introduction

Plasmodium falciparum are a unicellular protozoan organisms and toxic species that cause malaria in humans. It degrades hemoglobin in the acidic environment provided by the food vacuole [1]. When a female anopheles mosquito attacks human, malaria infection begins in the form of sporozoites into the bloodstream and its life cycle adopts many different stages [2]. These sporozoites are then quickly passed into the human liver where they upsurge exponentially into their cells to form merozoites. Merozoites attack red blood cells (erythrocytes) and again multiply until the cells burst to become trophozoites, schizonts, and gametocytes, during the last three stages of the anopheles life cycle as shown in Figure 1.

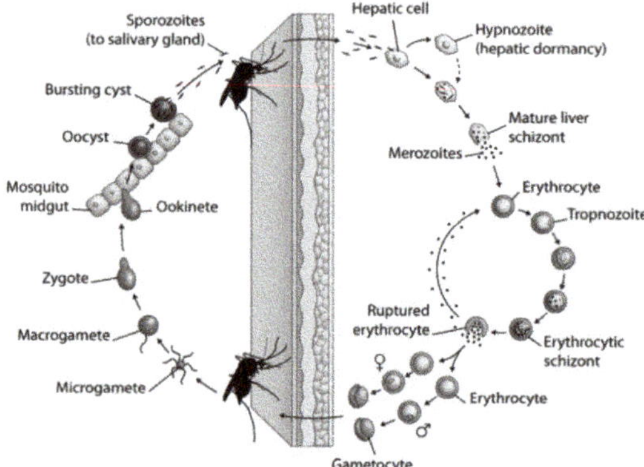

Figure 1. When a mosquito bites a human (host for malarial parasite) it causes infection by injecting sporozoites into the body, where it adversely affects hepatocyte's shape. Sporozoites grow rapidly in hepatocytes to become merozoites, while merozoites grow rapidly causing hepatocytes to burst and infect neighboring hepatocytes. When a mosquito bites the malaria patient the gametocytes that are produced from merozoites are taken by a mosquito. For the next 10 to 14 days the gametocytes produce sporozoites which are transferred to the saliva gland waiting for a mosquito to bite a healthy person and cause infection.

In eukaryotic cells, a mitochondrion is a membrane-bound organelle found in the cytoplasm which acts as a powerhouse of the cell responsible for cellular respiration and the production of adenosine triphosphate [3]. The inner membrane of cytoplasm comprises of different proteins such as enzymes, which are required for biochemical reactions. The mitochondrion has its own DNA (deoxyribonucleic acid) and ribosomes, which are 70 percent as that of prokaryote cells. In a cell, the mitochondrion is one of the important organelles which controls cellular metabolism and produces energy. Biologists have revealed that there are no significant similarities between mitochondrial proteins and human homologs [4].

Considering the constructive role of mitochondrial protein sequences in bioinformatics, proteomics, and cellular biology, many researcher's interest has been redirected to identify these biological sequences, but still it has been a challenging problem for them. With the invention of modern sequencing technologies, the number of these proteins has increased with rapid acceleration in the protein databanks. In 1990, only 3939 protein sequences were reported in the Uniprot database. According to the recent release statistics of protein databank, this number reached 550,000 in 2019 (11 December) [5].

There are two main approaches followed by researchers in the protein sequence prediction domain. The first category include machine learning-based approaches, where they employ features extractions methods in order to extract various patterns from biological sequences. Most of the existing literatures followed this approach. The second one is deep learning-based approach which extracts deep features and contextual information from proteins, which improves the prediction accuracy significantly. We briefly discuss the related works for both the mentioned approaches.

1.1. Machine Learning Approach towards Mitochondria Proteins Identification

In past decades, numerous machine learning algorithms and computational biological techniques are proposed for the categorization of mitochondrial and non-mitochondrial proteins via complex sequences. Bender et al. [6] evaluated MPPF by principal component analysis, statistical methods, and supervised neural network. They developed a model PlasMit based on extracting new composition patterns from proteins which efficiently predicted the mitochondrial proteins. R Verma et al. [7] combined two feature descriptors i.e., split amino acid and position specific scoring metrics, in order to predict mitochondrial proteins accurately. Jia et al. [8] considered mitochondrial proteins as attractive targets for anti-malarial drugs, but manual identification of these proteins is a difficult and time-consuming task. Therefore, they used two proteins encoding approaches such as bi-profile Bayes and split amino acid composition in order to extract specific pattern features from the amino acid chain. Authors trained a support vector machine classifier on these statistical features for final prediction. Afridi et al. [9] proposed genetic programming and an ensemble approach based on the feature extraction method for mitochondrial protein classification. Ding et al. [10] analyzed the variance for the accurate prediction of mitochondrial proteins. They suggested that combining more and more features is not a reliable approach because it takes more time in execution and contained redundant values which degraded the performance of the model. Therefore, to reduce the dimensionality and select the optimal features, researchers in this article used analysis of variance (ANOVA). Due to the complexity of the Plasmodium falciparum genome, the prediction of MPPF is more difficult than other species. Chen et al. [11] proposed an n-peptide composition of reduced amino acid alphabet which is obtained from a structural alphabet named protein as a feature parameter; the increment of diversity was firstly proposed to predict mitochondrial proteins. For instance, Cai et al. [12] applied support vector machine (SVM) based algorithm to train a predictor on three types of feature descriptors technique including transition (T), composition (C), and distribution (D) for obtaining additional physicochemical properties of different amino acids. R Kumar et al. [13] proposed a two-level model named as SubMitoPred. In the first level, authors predicted mitochondrial proteins while in the second level, they forecasted the sub-classes of mitochondrial localization. The whole model was based on the combination of SVM and the Pfam information domain. For further improvement, C Savojardo et al. [14] developed the deep learning model DeepMito. They trained and tested the model on a new high-quality dataset. Furthermore, they also developed a webserver for predicting mitochondrial and sub-mitochondrial localization. DNA-binding proteins (DBP's) can be used for the regulation of transcription and gene expression along with the identification of particular nucleotides. Therefore, for accurate and precise predictions of DBP's, Waris et al. [15] used evolutionary profiles position-specific scoring matrix for sequence encoding and a support vector machine for classification. Most of the available drugs are prepared to target the membrane proteins. Discriminating these proteins via computer vision techniques is an effective and timesaving as well. Therefore, Hayat et al. [16] through efforts of a computational method, accurately predicted the membrane proteins via different machine learning algorithms.

1.2. Deep Learning Approach towards Mitochondria Proteins Identification

Some researchers have utilized the full benefits of deep learning techniques and employed them to predict different types of protein sequences [17–19]. Delong et al. [20] attempted for the first time, to generate the original idea in deep learning for the discrimination of DNA binding proteins and non-DNA binding proteins. Qinhu et al. [21] improved the inherent weak supervision biological information prediction by proposing a new procedure established on CNN features with sequence-based learning to classify the DNA binding proteins. Recently, for sequencing learning, Qu et al. [17] applied a sequence learning network known as a recurrent neural network (RNN) with CNNs to predict those proteins which are attached to DNA. Compared to the traditional methods, deep learning techniques enhance the flexibility of extracted optimal features from sequences. It is not only the selection of a large number of proteins that are made possible for model training, but the process of speedy and accurate prediction is also enhanced. From earlier scholar's work, in addition to protein features, contextual information has also been observed as a valuable feature [22]. With an inspiration of this concept, if an amino acid sequence also suppresses contextual features, then it might increase the prediction score.

In this article, a CNN model is proposed by employing MBD-LSTM for Plasmodium mitochondria protein identification. In the first encoding layer, we assign an integer value to each amino acid in order to encode protein sequences and find the maximum length of the protein, and the next layer is the embedded layer where each word is converted to fix the length vector. Further, these matrices are passed through the CNN model containing three convolutional layers followed by max pooling layers. Finally, these deep features are fed into the MBD-LSTM layer for sequence learning and the last dense layer generates the optimal output. The proposed novel framework makes the following main contributions for the identification of Plasmodium falciparum parasite mitochondrial proteins.

- Considering the lack of effective vaccination, a rise in drug-resistant Plasmodium parasites, and the lethal nature of malaria, we propose a novel sequence-based framework MPPIF-Net to efficiently discriminate Plasmodium mitochondria and non-mitochondria proteins. The proposed model is useful in developing vaccines against malaria parasites.
- With the rising sequencing technology, the number of various proteins increases day by day with rapid acceleration in the protein databanks. In the aforementioned literature, researchers follow machine learning and computational techniques, which revealed inadequate performance while capturing contextual features from biological sequence patterns, yielding non-representative classifiers. In this study, we pursue a deep learning approach, which is capable of extracting contextual features and apply a sequence learning mechanism to efficiently classify the nature of proteins with the assistance of CNN and MBD-LSTM.
- Due to the unavailability of a large benchmark dataset of Plasmodium mitochondrial proteins, in this paper, we prepared a new dataset from the Uniprot site which contains both mitochondria and non-mitochondria proteins. The types of proteins mentioned in our dataset are passed from CD-Hit software to detect and remove similarity and short length proteins to optimally acquire a preprocessed and adoptable dataset.
- To validate the adoptability of our proposed model, we also made an extensive experimentation on the benchmark datasets, that is designed using mitochondrial proteins of another organism. The proposed model responded with convincing accuracy on this dataset, thereby validating the fact that our model is adoptable not only to the mitochondria proteins of the Plasmodium organism, but is trust-worthy to classify mitochondria proteins of other species as well.

The remaining article is divided into three further sections; Section 2 briefly explains the proposed system. Results and discussion are explained in Section 3 and in Section 4 we present the conclusions and future directions.

2. Proposed Methodology

Our proposed model mainly contains three modules which are further divided into five phases: (1) the preprocessing phase, in which we collect the sequence data from the protein databank and apply some techniques for the refining of data; (2) the encoding phase, in which we simply assign a natural number to each amino acid and also search the maximum length of the sequence; (3) the embedding phase, which is a mapping procedure in which individual words in the separate vocabulary will be inserted into a continuous vector space; (4) the convolution phase, in which we create one-dimensional vector from the encoded amino acid that is advanced to the CNN layers for deep features extraction; and finally, (5) the MBD-LSTM phase, where all the features are passed through this network for sequence learning to generate final output, as shown in Figure 2. The details of all phases are given in the subsequent sections. After passing the protein sequence (Pseq), the affinity scores of mitochondrial proteins are calculated by the Equation (1).

$$A(Pseq) = (Encoding\ (Embedding\ (CNN\ (MBD\text{-}LSTM)))) \tag{1}$$

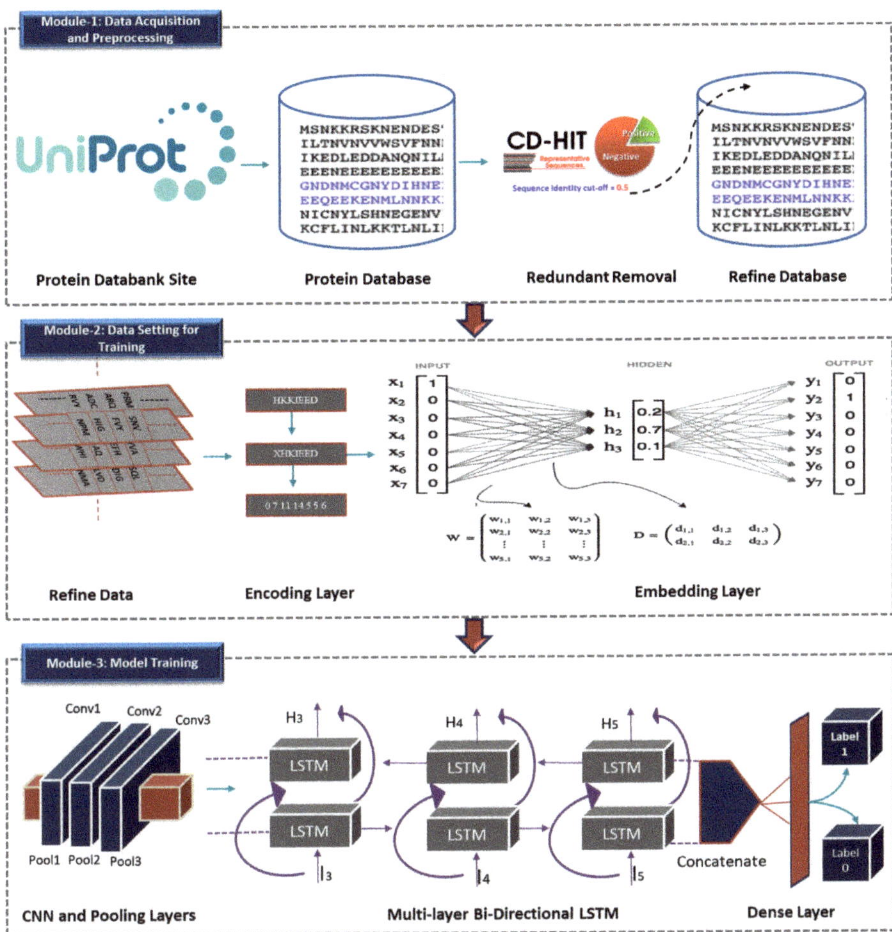

Figure 2. The proposed framework comprises of three modules. Module 1 describes the phenomena of sequence acquisition in which collection and preprocessing is channeled to eliminate redundancy. The polished sequences are forwarded to Module 2, where the alphabet was converted to natural numbers; after that we utilized embedding layers to generate fixed length vectors. Finally, in Module 3, we passed one-dimensional data to CNN deep contextual features extraction and then employed multi-layer bi-directional long short-term memory MBD-LSTM for sequence learning. Afterword, sigmoid activation is applied to predict final probability scores; either the output is related to mitochondrial (Label 1) or non-mitochondria (Label 0) which are then evaluated in terms of accuracy.

For predicting the output label data, we used the sigmoid activation function, and to evaluate the performance of the network, binary cross entropy is applied.

2.1. Raw Data Acquisition and Preprocessing

The biological sequences are obtained from the Uniprot protein databank in FASTA format, which is basically a text-based format. It is easily accessible, and downloadable protein sequences of any organism. In this study, we utilized three datasets such as PF2095, PF175, and MPD to estimate the model accuracy. Due to the unavailability of the massive number of MPPF, we collected 1701 raw sequences by searching the keywords "plasmodium falciparum mitochondrion", that are considered as positive samples. Furthermore, we collected 2075 negative samples by searching the "non-mitochondria proteins". After extracting the sequences, we inserted these sequences to CD-Hit software to removed sequences with 80% similarity and shorter length (less than 40) of amino acids. After the refining process, we collected 890 and 1205 as positive and negative samples for classification purposes. The remaining two benchmark datasets were publicly available, so there is no need to pass it from the preprocessing phase.

2.2. Encoding Protein Sequences

In machine learning, various feature extraction approaches are proposed, such as amino acid composition, dipeptide composition, split amino acid composition, pseudo amino acid composition, position specific scoring matrices, and n-gram methods, etc. [18]. These manual protein encoding techniques usually extract low-level features which sometimes degrade the machine learning model, mostly in classification problems. Nowadays, a renowned approach for the achievement of a high success rate is deep learning mechanism, which is the subset of machine learning in artificial intelligence. Here we simply allocate a natural number to each amino acid [23]. For instance, we have a protein sequence like 'AKILMEF', so the encoding of each amino acid is represented as (1, 4, 5, 8, 9, 10, 11). Symbol along with code for each amino acid is shown in Table 1. This encoding is efficient as compared to the sparse vector. An important aspect for consideration while assigning numbers to amino acids is related to order, which states that order does not affect a model's performance at all.

Table 1. Symbolic representation of each Amino Acids and its code.

Amino Acids	Letters	Code
Alanine	A	1
Cysteine	C	2
Aspartic	D	3
Glutamic	E	4
Phenylalanine	F	5
Glycine	E	6
Histidine	H	7
Isoleucine	I	8
Lysine	K	9
Leucine	L	10
Methionine	M	11
Asparagine	N	12
Proline	P	13
Glutamine	Q	14
Arginine	R	15
Serine	S	16
Threonine	T	17
Valine	V	18
Tryptophan	W	19
Tyrosine	Y	20

The encoding phase only creates a digital vector of a proteins sequence with variable length. First, we find the maximum length of protein in a dataset. In this paper, the max-length vector value is (5253, 1280, and 1402) for distinct datasets which usually depends upon the sequences in the dataset. As we already know, protein sequences usually exhibit different lengths while in deep learning, we are required to keep a fixed length for all protein sequences. For example, for a protein whose length vector is smaller than the maximum-length vector, a unique value zero is placed, at the end in order to keep the same alignment of all the sequences. An encoding of protein example is given in Equation (2).

$$\text{Protein Seq1} = \text{Encoding (Sequence)} = (11, 12, 16, 17, 0) \tag{2}$$

2.3. Embedding Layer

It is a very difficult task to encode each word manually. This layer gives us an automatic and efficient way of representing words or documents in which matching words have a similar encoding [16]. This work is done by just multiplying one hot vector from the left with a weight matrix $W \in R^{d\,*|V|}$ where $|V|$ represents the number of primary symbols related to the vocabulary as shown in Equation (3).

$$V_Z = W_{X_t} \tag{3}$$

As a result, the input sequence of amino acids becomes a solid valued vector ($z = 1, 2, 3, 4, \ldots$ n). In the embedding layer, assume that the output dense vector length is 8, and each number map corresponds to a fixed vector length. After passing through layer proteins, the sequence becomes an 8×8 matrix e.g., as exposed in Equation (4). We may represent Thyronine amino acid with [0.5, −0.8, 0.7, 0.4, 0.3, −0.5, −0.7, 0.8] and Methionine with [0.4, −0.4, 0.5, 0.6, 0.2, −0.1, −0.3, 0.2].

$$\text{ProteinSeq2} = \begin{pmatrix} 0.1 & -0.4 & 0.1 & 0.2 & 0.6 & 0.4 & -0.1 & 0.1 \\ 0.4 & -0.4 & 0.5 & 0.6 & 0.2 & -0.1 & -0.3 & 0.2 \\ 0.2 & -0.2 & 0.6 & 0.7 & -0.1 & 0.1 & -0.2 & 0.1 \\ 0.5 & -0.2 & 0.1 & 0.6 & 0.2 & -0.6 & -0.2 & 0.9 \\ 0.4 & -0.4 & 0.5 & 0.6 & 0.2 & -0.1 & -0.3 & 0.2 \\ 0.8 & -0.5 & 0.4 & 0.7 & 0.5 & -0.2 & -0.5 & 0.3 \\ 0.9 & -0.6 & 0.7 & 0.8 & 0.2 & -0.1 & -0.2 & 0.7 \\ 0.5 & -0.8 & 0.7 & 0.4 & 0.3 & -0.5 & -0.7 & 0.8 \end{pmatrix} \tag{4}$$

2.4. Convolution Layer

The deep learning model works very efficiently in image processing and video analysis by extracting the deep features based on convolutional and pooling layers [24,25]. In case of images or videos, we directly give input data to the model because their data already exhibits a matrix arrangement [26]. While working with protein sequences, first, we prepare data in the form of a matrix with fixed-size and forwards to the convolution layer for processing like images. In this work, the model exhibits three convolution layers and each one is followed by a max pooling layer for deep features extraction. In this layer, we use 3×8 filters to scan the protein seq2 and obtain a new feature map as shown in Equations (5) and (6).

$$\textbf{Filter} = \begin{bmatrix} 0.2 & 0.2 & -0.3 & 0.8 & 0.5 & 0.3 & 0.2 & -0.2 \\ 0.1 & 0.3 & -0.3 & 0.6 & 0.1 & 0.3 & -0.2 & 0.3 \\ 0.8 & -0.2 & 0.3 & -0.5 & 0.6 & 0.3 & 0.2 & 0.1 \end{bmatrix} \tag{5}$$

$$\text{Protein seq3} = \text{Convolution (Seq2)} = \begin{bmatrix} 0.48 \\ 0.53 \\ 0.75 \\ 0.20 \\ 0.25 \\ 0.62 \\ 0.40 \end{bmatrix} \tag{6}$$

In max pooling layer, the sliding window takes the highest value of the two numbers as shown in Equation (7)

$$\text{Protein seq4} = \text{Max Pooling (Seq3)} = \begin{bmatrix} 0.65 \\ 0.53 \\ 0.48 \\ 0.62 \end{bmatrix} \tag{7}$$

2.5. MBD-LSTM Layer

For the complications and issues related to short-term memory sequencing, RNN is employed, mostly for the cases where a long sequence is required to handle and stored for both forward and backward steps. As a result of the continuation of this procedure, RNN may depart crucial information out of the initial sequence data. But during backpropagation, a vanishing gradient issue is encountered, that makes it hard to memorize long-term changes in sequence [27,28]. Throughout the propagation process, the neural network weights are updated and shrink due to the gradient. These extremely minor weights do not participate to the learning process in an RNN, and also layers stop learning due to acquiring such a small gradient. In this situation the RNN does not have a capability to store longer sequence modifications that were observed previously. LSTM provides a solution by incorporating a short-term memory unit which is a special recurrent neural network architecture. LSTM emphases build memory cells and gates that regulate to process and store information and also allow when to update and forget the hidden states of the network [29]. The internal structure review of LSTM contains memory cell state S_{st-1}. These cells directly relate to H_{st}-1 which is the middle output state, and the successive state X_{st} controls the internal state vector which is required to be upgraded. There are three gates in LSTM structure; input gates N_{st}, forget gates F_{st}, and the output gate O_{st}. The mathematical notation of these gates are as follows.

$$F_{ST} = \sigma(W_{FX}X_{ST} + W_{FH}H_{ST-1} + B_F) \tag{8}$$

$$I_{ST} = \sigma(W_{IX}X_{ST} + W_{IH}H_{ST-1} + B_I) \tag{9}$$

$$N_{ST} = \phi(W_{NX}X_{ST} + W_{NH}H_{ST-1} + B_N) \tag{10}$$

$$O_{ST} = \sigma(W_{OX}X_{ST} + W_{OH}H_{ST-1} + B_O) \tag{11}$$

$$S_{ST} = \sigma(G_T\theta I_{ST} + S_{ST-1}\,\theta\,B_S) \tag{12}$$

$$H_{ST} = \phi(S_{ST-1})\,\theta O_{ST} \tag{13}$$

In Equations (8)–(13), the network inputs weight matrices are represented by $W_{FX}, W_{FH}, W_{IX}, W_{IH}, W_{NX}, W_{NH}, W_{OX}$, and W_{HO}. Here, θ is used for the multiplication in an elementwise manner. The two activation functions such as sigmoid and tanh are represented by σ and ϕ. The single time step of LSTM architecture is shown in Figure 3a. In this article, we evaluated the performance of MBD-LSTM for protein sequence identification. The idea of a MBD-LSTM is developed from traditional bidirectional RNN [30], which also processes the hidden layer input sequence data in both forward and backward direction. MBD-LSTM has achieved significant results in speech recognition [31], summarization [32], classification, energy consumption prediction [33], and text generation. The structure of MBD-LSTM consists of forward and backward layers as shown in Figure 3b. The output of the forward layer $I_T^{>}$ is analyzed through input data from T − n to T − 1, while the output data of the backward layer $H_T^{<}$ is generated through reversed inputs such as from T − n to T − 1. Final MBD-LSTM generates the O_T output vector as illustrated in equation (14).

$$O_T = \sigma(I_T^{>},\ H_T^{>}) \tag{14}$$

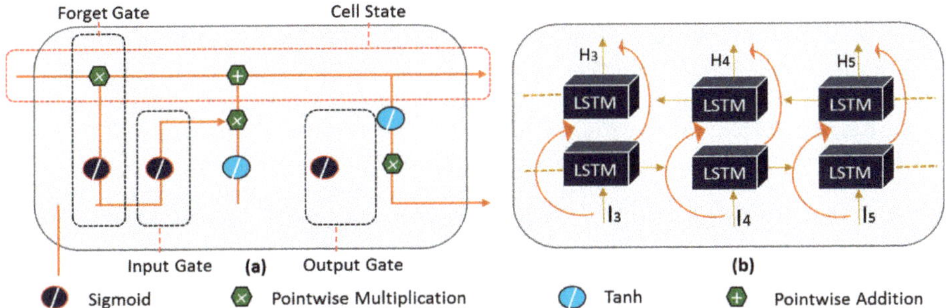

Figure 3. (a) Internal architecture of LSTM, comprising of multiple gates along with LSTM cells for stimulation of numerous operations, permitting the gates to store and omit related information; (b) MBD-LSTM, which acquires input data sequences and then proceeds in a forward and backward direction.

In Equation (11), σ combines an output sequence of two layers, which is also known as summation function.

3. Results and Discussion

In this portion, an in-depth analysis over comprehensive experiments which are performed on three protein sequence datasets and detailed discussion of comparative studies of the proposed model with state-of-the-art techniques is presented.

3.1. Datasets

The benchmark dataset of Plasmodium mitochondrial protein sequences was obtained from [5]. The total number of proteins in dataset is 175, which contains 40 positive (mitochondrial proteins) and 135 negative (non-mitochondrial proteins) samples. We represent this dataset by Plasmodium falciparum 175 (PF175). The second raw dataset (PF2095) was obtained from the universal protein resource (Uniprot) which contains 890 positive samples and 1205 negative samples.

Similarly, a third dataset was downloaded from [34] which contained 499 positive samples. In this paper, we denote this dataset by (MPD) as shown in Table 2. Actually, 2833 proteins were obtained from the protein databank site known as Swiss-Prot by searching the keyword mitochondrial. Afterward, those proteins were then excluded with ambiguous words, such as SIMILARITY, POTENTIAL, or PROBABLE and FRAGMENTS. Furthermore, 681 proteins were collected belonging to locations other than mitochondrial site.

Table 2. Plasmodium falciparum mitochondria protein datasets.

Dataset	Positive Sample	Negative Sample
PF175	40	135
PF2095	890	1205
MPD	499	250

By applying the preprocessing and eliminating the ambiguous data, we selected 250 proteins as non-mitochondrial.

3.2. Experimental Setup

Using two benchmark and one own prepared protein datasets, we analyzed and verified the efficiency of the proposed model. The model was trained on a Titan Intel Core i5-6600 processor with X (Pascal)/PCLe/SSE2 GPU, having 64GB of memory using 16.4 LTS Ubuntu operating system.

The proposed deep learning model was executed in version 3.5 of python, version 2.2.4 of Keras, and version 1.12 of TensorFlow backend along with an Adam employed as an optimizer. To find the most favorable selection of the hyperparameter of each model, several experiments were conducted. At last, we selected 50 epochs to train the model with a batch size of 100. The PF2095 and MPD samples were split into training 70% and testing 30%, and due to fewer numbers of protein samples in PF175 we kept 80% data in training, and 20% data is utilized for model evaluation.

3.3. Evaluation Metrics

In this study, a couple of assessment measures are used for the evaluation of the proposed model. These parameters include accuracy, sensitivity, and specificity. The mathematical formulas are defined in the Equation (15)–(17).

$$\text{Accuracy} = \frac{\text{TP} + \text{TN}}{\text{TP} + \text{TN} + \text{FP} + \text{FN}} \times 100 \tag{15}$$

$$\text{Sensitivity} = \frac{\text{TP}}{\text{TP} + \text{FN}} \times 100 \tag{16}$$

$$\text{Specificity} = \frac{\text{TN}}{\text{TN} + \text{FP}} \times 100 \tag{17}$$

Now, let us assume that the mitochondrial protein is positive, and the non-mitochondrial protein is negative. The true positive (TP) is that value in which predictive and actual value is positive and true negative (TN) is the value in which predicted value and actual value is negative. Similarly, the false positive (FP) is the value in which a machine predicted as positive but actually it is a negative value and the false negative (FN) is that value in which machine predicted as negative class but actually it is related to positive class value.

3.4. Ablation Study on PF2095

In this subsection, we conduct an ablation study after comprehensive experiments to analyze the three models in terms of accuracy, sensitivity, and specificity on the PF2095 dataset, which is a new mitochondria proteins of Plasmodium dataset comprising 890 positive and 1205 negative samples. In this dataset 70% of total samples are set for training, and the remaining 30% are used for model evaluation. First, we perform our experiments on CNN-GRU which achieved 89.7% training accuracy, 88.0% testing, 90.4% sensitivity, and 88.9% specificity. The next model CNN-LSTM showed better performance compared to previous one. It obtained 93.5% training, 91.2% testing accuracy, 90.6% sensitivity, and 91.7% specificity. The proposed model MPPIF-NET used CNN with integration of MBD-LSTM with the same number of parameters. It is experimentally proved that the last hybrid approach shows supremacy of performance which obtained 98.2% training accuracy, 97.6% is testing performance of the model, 98.1% of its sensitivity, and 97.2% specificity. The detailed experimental evaluation results are depicted in Table 3 and confusion metrics of the MPPIF-NET are shown in Figure 4.

Table 3. Training and testing performance of the MPPIF-NET on different models and datasets.

Dataset	Model	Training Accuracy	Testing Accuracy	Sensitivity	Specificity
PF2095	CNN-GRU	89.7	88.0	90.4	88.9
	CNN-LSTM	93.5	91.2	90.6	91.7
	CNN-MBD-LSTM (Proposed)	98.2	97.6	98.1	97.2
PF175	CNN-MBD-LSTM (Proposed)	100	97.1	100	96.2
MPD	CNN-MBD-LSTM (Proposed)	99.7	99.5	99.3	100

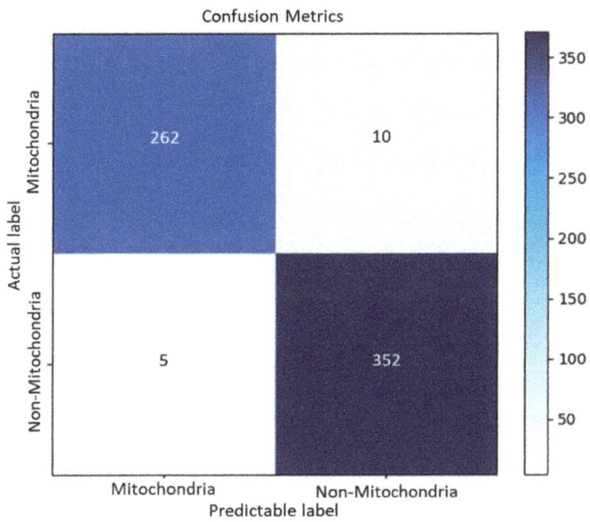

Figure 4. Confusion metrics of the proposed method over the PF2095 dataset.

3.5. Experimental Evaluation on PF175

We used irregular data in our experiments along with a hold-out technique which is the simplest kind of cross validation. The data was divided into training and testing. We trained our proposed model on 80% of the data and the remaining 20% of data were used for evaluation purposes. During experiments we updated different parameters to achieve good performance. After numerous experiments we set these parameters and their value; for example, maximum length of the protein is 1280 which depends upon the dataset, maximum features = 26, embedding size = 8, number of filters in convolutional are 32, pooling length = 2, batch size = 100, dropout = 0.2, and number of epochs is 50. We also checked different numbers of epochs and finally realized that the trained model fits the protein sequences well and predicts accurately on epochs 50.

Our model achieved better performance in terms of 100% training accuracy, 100% sensitivity, 96.2% specificity, and testing accuracy of 97.14%, which is higher than other state-of-the-art approaches. The confusion matrics of correctly and incorrectly predicted proteins are shown in Figure 5.

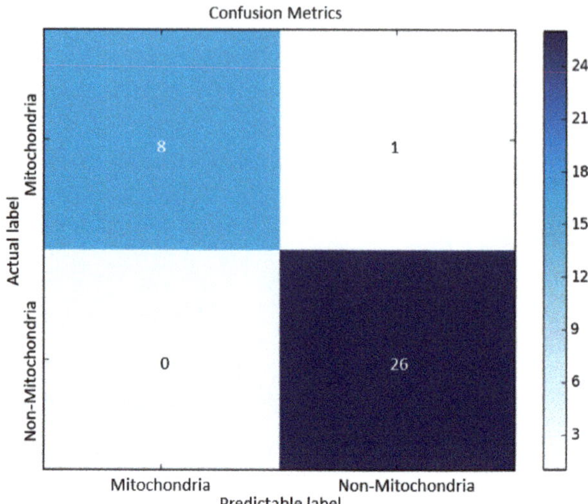

Figure 5. Confusion metrics of the proposed method over the PF175 dataset.

3.6. Experimental Evaluation on MPD

This dataset is also an unbalanced dataset and used the hold-out method during experiments. The data is divided into training and testing, which is 70% and 30%. For this dataset we also set the same parameters, except the maximum length of the protein which is 1402; maximum features = 26, embedding size = 8, number of filters in convolutional are 32, pooling length = 2, batch size = 100, dropout = 0.2 and the number of epochs is 50. We have done a lot of experiments with different setup parameters, but finally on epochs 50 we achieved better performance. The confusion matrics of correctly and incorrectly predicted proteins are shown in Figure 6.

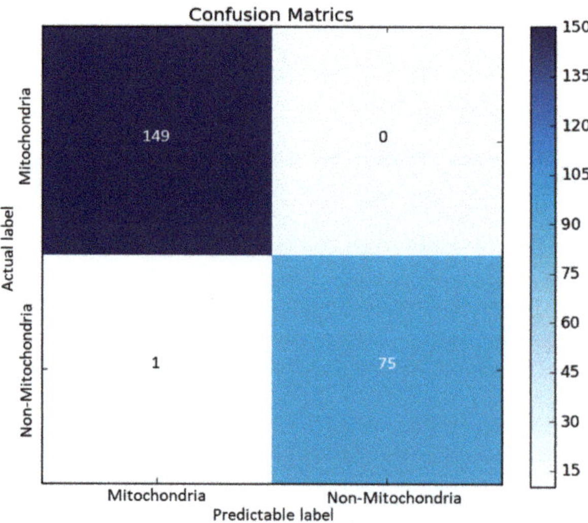

Figure 6. Confusion metrics of the proposed method over the mitochondria protein dataset (MPD) dataset.

Our model got 99.7% training accuracy, 99.3% sensitivity, 100% specificity, and a testing accuracy of 99.5%, which shows that the proposed model is superior in contrast to state-of-the-art techniques.

3.7. Comparative Analysis of the MPPIF-NET with Other Models on PF175

In the post genomic era, functional annotation is one of the major challenges. From the last decade, a vast number of machine learning and bioinformatic techniques have been proposed to predict protein functionality. The statistics of sequences are boosted day by day in the protein databanks. Identification of these biological sequences via laboratory methods was a laborious task. Therefore, we proposed a deep learning model for the accurate prediction of a huge number of proteins. Hence, it is important to evaluate the performance of models in order to compute the realistic performance of the model. For this we compared our proposed model with the state-of-the-art method using the same dataset. In the first attempt Bhasin et al. [35] proposed a model for eukaryotic subcellular localization protein prediction called (Eslpred) using a hybrid approach containing a dipeptide composition and PSI-BLAST. They achieved 69.71% accuracy, 73.33% specificity, and 57.50% sensitivity. Guda et al. [36] developed a new method for genome-scale prediction of the target mitochondria protein based on the composition of the amino acid and the occurrence frequency of each pattern which repeats in sequences. They achieved 80% accuracy, 87.41% specificity, and 55% sensitivity. Bender et al. [6] built a neural network model for the precise prediction of mitochondrial transit peptides which causes malaria. Due to the complex genomic sequence of PF, Chen et al. [11] developed the increment of diversity model in which a reduced amino acid composition was used in order to extract local features from the biological sequence. The prediction performance achieve 100% superior sensitivity rate, 89% specificity, and 92% accuracy as shown in Table (4). Mitochondria are vital organelles of eukaryotic cells which are involved in processing cellular death and human diseases; therefore, Afridi et al. [9] proposed an ensemble model known as Mito-GSAAC in which the main purpose was to examine an effective feature extraction approach. They achieved the highest specificity score of 95.56%, 93.21% accuracy, and 87.5% sensitivity. Accurate identification of the mitochondrial protein of Plasmodium falciparum is an essential role in the discovery of anti-malarial drug targets. Ding et al. [10] used a dipeptide composition for protein encoding. They also used the analysis of variance to overcome the issue of overfitting. They attained 97.1% accuracy, 90% sensitivity, and 99.3% specificity. The aforementioned state-of-the-art techniques utilized the machine learning approaches for the protein sequences prediction. We proposed a deep learning strategy for identification of these biological sequences which gave 97.14% superior testing accuracy compared to other discussed methods as shown in Table 4.

Table 4. MPPIF-NET comparative analysis with other models on PF175 dataset.

Method	Sensitivity	Specificity	Accuracy
Eslpred [34]	57.50	73.33	69.71
Mitopred [35]	55.00	87.41	80.00
PlasMit [5]	94.00	89.00	90.00
ID [10]	100	89.63	92.00
MitoGSAAC [8]	87.5	95.56	93.21
ANOVA [9]	90.0	**99.3**	97.1
MPPFI-Net	**100**	<u>96.2</u>	**97.14**

3.8. Comparative Analysis of the MPPIF-NET with Other Models on MPD

Mitochondria are the center and powerhouse of the eukaryotic cells. Pharmaceutical companies still desire such a system which accurately predicts the mitochondria protein of Plasmodium in order to prepare drugs. Therefore, Tan et al. [34] proposed an algorithm in order to evaluate the pair composition of amino acids. The extracted features are then passed to the support vector machine classifier for prediction of Plasmodium mitochondria proteins. The SVM model was evaluated which achieved 85% accuracy, 89.28% specificity, and 79.16% sensitivity. Jiang et al. [37] developed a new sequence-based method which is known as the Discrete Wavelet Transform for sequence prediction. They achieved 50.30% sensitivity, 95.74% specificity, and 76.53% accuracy. Afridi et al. [9] used four computational methods such as AAC, DPC, SAAC, and PAAC. Furthermore, they also evaluated the six machine learning algorithms, such as support vector machine, random forest, multilayer perceptron, AdaBoost, and bagging. Finally, on the basis of the ensemble classifier they achieved 92.62% accuracy, 91.52% specificity, and 90.96% sensitivity. Our proposed model performs well compared to the state-of-the-art methods, having 99.5% accuracy, 100% specificity, and 99.33% sensitivity as shown in Table 5.

Table 5. MPPIF-NET comparative analysis with other models on the MPD dataset.

Method	Sensitivity	Specificity	Accuracy
SVM 84-D [33]	79.16	89.28	85.00
DWT [36]	50.30	95.74	76.53
MitoGSAAC [8]	90.96	91.52	92.62
MPPFI-Net	**100**	**99.33**	**99.5**

4. Conclusions and Future Directions

For the identification of mitochondria proteins of Plasmodium some biologists are still concentrating on extracting new patterns from biological sequences and are searching for appropriate machine learning algorithms which accurately classify proteins. In this study, we proposed a deep leaning framework MPPFI-Net which is capable of extracting deep features automatically and can discriminate proteins quickly and accurately. We merged the CNN and MBD-LSTM in order to extract the contextual information from amino acids. Later on, we compared MPPFI-Net performance with the state-of-the-art models, and we conclude that the proposed framework speeds up the performance regarding both prediction accuracy and fitting uncharacterized data. In future, we will boost this work by fusing the traditional features and deep features.

Author Contributions: Conceptualization, S.U.K. and R.B.; methodology, S.U.K.; writing—original draft preparation, S.U.K.; review and editing, S.U.K. and R.B.; supervision, R.B. All authors have read and agreed to the published version of the manuscript.

Funding: The publication fees were supported by Prof. Ran Baik (HONAM University-20190125).

Acknowledgments: The author(s) disclosed receipt of the following financial support for the research, authorship, and/or publication of this article: this study was supported by the research fund from Honam University (2019).

Conflicts of Interest: The authors declare no conflict of interest.

References

1. Gazanion, E.; Vergnes, B. Protozoan parasite auxotrophies and metabolic dependencies. In *Metabolic Interaction in Infection*; Springer: Berlin/Heidelberg, Germany, 2018; pp. 351–375.
2. Dundas, K.; Shears, M.J.; Sinnis, P.; Wright, G.J. Important extracellular interactions between Plasmodium sporozoites and host cells required for infection. *Trends Parasitol.* **2019**, *35*, 129–139. [CrossRef] [PubMed]
3. Hou, X.S.; Wang, H.S.; Mugaka, B.P.; Yang, G.J.; Ding, Y. Mitochondria: Promising organelle targets for cancer diagnosis and treatment. *Biomater. Sci.* **2018**, *6*, 2786–2797. [CrossRef]
4. Devine, M.J.; Kittler, J.T. Mitochondria at the neuronal presynapse in health and disease. *Nat. Rev. Neurosci.* **2018**, *19*, 63. [CrossRef] [PubMed]
5. UniProtKB/Swiss-Prot UniProt 2019. Available online: https://www.uniprot.org/statistics/Swiss-Prot%202019_06 (accessed on 20 May 2020).
6. Bender, A.; van Dooren, G.G.; Ralph, S.A.; McFadden, G.I.; Schneider, G. Properties and prediction of mitochondrial transit peptides from Plasmodium falciparum. *Mol. Biochem. Parasitol.* **2003**, *132*, 59–66. [CrossRef] [PubMed]
7. Verma, R.; Varshney, G.C.; Raghava, G.P.S. Prediction of mitochondrial proteins of malaria parasite using split amino acid composition and PSSM profile. *Amino Acids* **2010**, *39*, 101–110.

8. Jia, C.; Liu, T.; Chang, A.K.; Zhai, Y. Prediction of mitochondrial proteins of malaria parasite using bi-profile Bayes feature extraction. *Biochimie* **2011**, *93*, 778–782. [CrossRef] [PubMed]

9. Afridi, T.H.; Khan, A.; Lee, Y.S. Mito-GSAAC: Mitochondria prediction using genetic ensemble classifier and split amino acid composition. *Amino Acids* **2012**, *42*, 1443–1454. [CrossRef] [PubMed]

10. Ding, H.; Li, D. Identification of mitochondrial proteins of malaria parasite using analysis of variance. *Amino Acids* **2015**, *47*, 329–333. [CrossRef] [PubMed]

11. Chen, Y.-L.; Li, Q.-Z.; Zhang, L.-Q. Using increment of diversity to predict mitochondrial proteins of malaria parasite: Integrating pseudo-amino acid composition and structural alphabet. *Amino Acids* **2012**, *42*, 1309–1316. [CrossRef] [PubMed]

12. Cai, C.Z.; Han, L.Y.; Ji, Z.L.; Chen, X.; Chen, Y.Z. SVM-Prot: Web-based support vector machine software for functional classification of a protein from its primary sequence. *Nucleic Acids Res.* **2003**, *31*, 3692–3697. [CrossRef]

13. Kumar, R.; Kumari, B.; Kumar, M. Proteome-wide prediction and annotation of mitochondrial and sub-mitochondrial proteins by incorporating domain information. *Mitochondrion* **2018**, *42*, 11–22. [CrossRef] [PubMed]

14. Savojardo, C.; Bruciaferri, N.; Tartari, G.; Martelli, P.L.; Casadio, R. DeepMito: Accurate prediction of protein sub-mitochondrial localization using convolutional neural networks. *Bioinformatics* **2020**, *36*, 56–64. [CrossRef] [PubMed]

15. Waris, M.; Ahmad, K.; Kabir, M.; Hayat, M. Identification of DNA binding proteins using evolutionary profiles position specific scoring matrix. *Neurocomputing* **2016**, *199*, 154–162. [CrossRef]

16. Hayat, M.; Khan, A. MemHyb: Predicting membrane protein types by hybridizing SAAC and PSSM. *J. Theor. Biol.* **2012**, *292*, 93–102.

17. Qu, Y.H.; Yu, H.; Gong, X.J.; Xu, J.H.; Lee, H.S. On the prediction of DNA-binding proteins only from primary sequences: A deep learning approach. *PLoS ONE* **2017**, *12*, e0188129. [CrossRef]

18. Zeng, H.; Edwards, M.D.; Liu, G.; Gifford, D.K. Convolutional neural network architectures for predicting DNA–protein binding. *Bioinformatics* **2016**, *32*, i121–i127. [CrossRef] [PubMed]

19. Qiu, W.; Li, S.; Cui, X.; Yu, Z.; Wang, M.; Du, J.; Peng, Y.; Yu, B. Predicting protein submitochondrial locations by incorporating the pseudo-position specific scoring matrix into the general Chou's pseudo-amino acid composition. *J. Theor. Boil.* **2018**, *450*, 86–103. [CrossRef] [PubMed]

20. Alipanahi, B.; Delong, A.; Weirauch, M.T.; Frey, B.J. Predicting the sequence specificities of DNA-and RNA-binding proteins by deep learning. *Nat. Biotechnol.* **2015**, *33*, 831–838. [CrossRef] [PubMed]

21. Zhang, Q.; Zhu, L.; Bao, W.; Huang, D.S. Weakly-supervised convolutional neural network architecture for predicting protein-DNA binding. *IEEE/ACM Trans. Comput. Biol. Bioinform.* **2018**. [CrossRef]

22. Melamud, O.; Goldberger, J.; Dagan, I. context2vec: Learning generic context embedding with bidirectional lstm. In Proceedings of the 20th SIGNLL Conference on Computational Natural Language Learning, Berlin, Germany, 7–12 August 2016.

23. Monteiro, N.R.; Ribeiro, B.; Arrais, J.P. Deep Neural Network Architecture for Drug-Target Interaction Prediction. In Proceedings of the International Conference on Artificial Neural Networks, Munich, Germany, 17–19 September 2019; Springer: Cham, Switzerland, 2019.

24. Khan, S.U.; Haq, I.U.; Rho, S.; Baik, S.W.; Lee, M.Y. Cover the Violence: A Novel Deep-Learning-Based Approach towards Violence-Detection in Movies. *Appl. Sci.* **2019**, *9*, 4963. [CrossRef]

25. Haq, I.U.; Muhammad, K.; Ullah, A.; Baik, S.W. DeepStar: Detecting starring characters in movies. *IEEE Access* **2019**, *7*, 9265–9272. [CrossRef]

26. Ullah, A.; Muhammad, K.; Del Ser, J.; Baik, S.W.; de Albuquerque, V.H.C. Activity recognition using temporal optical flow convolutional features and multilayer LSTM. *IEEE Trans. Ind. Electron.* **2018**, *66*, 9692–9702. [CrossRef]

27. Bengio, Y.; Simard, P.; Frasconi, P. Learning long-term dependencies with gradient descent is difficult. *IEEE Trans. Neural Netw.* **1994**, *5*, 157–166. [CrossRef] [PubMed]

28. Hochreiter, S.; Bengio, Y.; Frasconi, P.; Schmidhuber, J. Gradient flow in recurrent nets: The difficulty of learning long-term dependencies. In *A Field Guide to Dynamical Recurrent Neural Networks*; IEEE Press: Linz, Austria, 2001.

29. Hochreiter, S.; Schmidhuber, J. Long short-term memory. *Neural Comput.* **1997**, *9*, 1735–1780. [CrossRef] [PubMed]

30. Schuster, M.; Paliwal, K.K. Bidirectional recurrent neural networks. *IEEE Trans. Signal Process.* **1997**, *45*, 2673–2681.

31. Kwon, S. A CNN-Assisted Enhanced Audio Signal Processing for Speech Emotion Recognition. *Sensors* **2020**, *20*, 183.

32. Hussain, T.; Muhammad, K.; Ullah, A.; Cao, Z.; Baik, S.W.; de Albuquerque, V.H.C. Cloud-Assisted Multiview Video Summarization Using CNN and Bidirectional LSTM. *IEEE Trans. Ind. Inform.* **2019**, *16*, 77–86. [CrossRef]

33. Ullah, F.U.M.; Ullah, A.; Haq, I.U.; Rho, S.; Baik, S.W. Short-Term Prediction of Residential Power Energy Consumption via CNN and Multilayer Bi-directional LSTM Networks. *IEEE Access* **2019**. [CrossRef]

34. Tan, F.; Feng, X.; Fang, Z.; Li, M.; Guo, Y.; Jiang, L. Prediction of mitochondrial proteins based on genetic algorithm–partial least squares and support vector machine. *Amino Acids* **2007**, *33*, 669–675. [CrossRef]

35. Bhasin, M.; Raghava, G. ESLpred: SVM-based method for subcellular localization of eukaryotic proteins using dipeptide composition and PSI-BLAST. *Nucleic Acids Res.* **2004**, *32* (suppl. 2), W414–W419. [CrossRef]

36. Guda, C.; Fahy, E.; Subramaniam, S. MITOPRED: A genome-scale method for prediction of nucleus-encoded mitochondrial proteins. *Bioinformatics* **2004**, *20*, 1785–1794. [CrossRef] [PubMed]

37. Jiang, L.; Li, M.; Wen, Z.; Wang, K.; Diao, Y. Prediction of mitochondrial proteins using discrete wavelet transform. *Protein J.* **2006**, *25*, 241–249. [CrossRef] [PubMed]

Article

Measuring Performance Metrics of Machine Learning Algorithms for Detecting and Classifying Transposable Elements

Simon Orozco-Arias [1,2,*], **Johan S. Piña** [3], **Reinel Tabares-Soto** [4], **Luis F. Castillo-Ossa** [2], **Romain Guyot** [4,5] **and Gustavo Isaza** [2,*]

1   Department of Computer Science, Universidad Autónoma de Manizales, Manizales 170001, Colombia
2   Department of Systems and Informatics, Universidad de Caldas, Manizales 170004, Colombia;
    luis.castillo@ucaldas.edu.co
3   Research Group in Software Engineering, Universidad Autónoma de Manizales, Manizales 170001,
    Colombia; johan.pinad@autonoma.edu.co
4   Department of Electronics and Automation, Universidad Autónoma de Manizales, Manizales 170001,
    Colombia; rtabares@autonoma.edu.co (R.T.-S.); romain.guyot@ird.fr (R.G.)
5   Institut de Recherche pour le Développement, Univ. Montpellier, UMR DIADE, 34394 Montpellier, France
*   Correspondence: simon.orozco.arias@gmail.com (S.O.-A.); gustavo.isaza@ucaldas.edu.co (G.I.)

Received: 25 April 2020; Accepted: 22 May 2020; Published: 27 May 2020

Abstract: Because of the promising results obtained by machine learning (ML) approaches in several fields, every day is more common, the utilization of ML to solve problems in bioinformatics. In genomics, a current issue is to detect and classify transposable elements (TEs) because of the tedious tasks involved in bioinformatics methods. Thus, ML was recently evaluated for TE datasets, demonstrating better results than bioinformatics applications. A crucial step for ML approaches is the selection of metrics that measure the realistic performance of algorithms. Each metric has specific characteristics and measures properties that may be different from the predicted results. Although the most commonly used way to compare measures is by using empirical analysis, a non-result-based methodology has been proposed, called measure invariance properties. These properties are calculated on the basis of whether a given measure changes its value under certain modifications in the confusion matrix, giving comparative parameters independent of the datasets. Measure invariance properties make metrics more or less informative, particularly on unbalanced, monomodal, or multimodal negative class datasets and for real or simulated datasets. Although several studies applied ML to detect and classify TEs, there are no works evaluating performance metrics in TE tasks. Here, we analyzed 26 different metrics utilized in binary, multiclass, and hierarchical classifications, through bibliographic sources, and their invariance properties. Then, we corroborated our findings utilizing freely available TE datasets and commonly used ML algorithms. Based on our analysis, the most suitable metrics for TE tasks must be stable, even using highly unbalanced datasets, multimodal negative class, and training datasets with errors or outliers. Based on these parameters, we conclude that the F1-score and the area under the precision-recall curve are the most informative metrics since they are calculated based on other metrics, providing insight into the development of an ML application.

Keywords: transposable elements; metrics; machine learning; deep learning; detection; classification

---

1. Introduction

Transposable elements (TEs) are genomic units able to move within and among the genomes of virtually all organisms [1]. They are the main contributors to genomic diversity and genome size variations [2], except for of polyploidy events. Also, TEs perform key genomic functions involved in chromosome structuring, gene expression regulation and alteration, adaptation and evolution [3], and centromere composition in plants [4]. Currently, an important issue in genome sequence analyses is to rapidly identify and reliably annotate TEs. However, there are major obstacles and challenges in the analysis of these mobile elements [5], including their repetitive nature, structural polymorphism, species specificity, as well as high divergence rate, even across close relative species [6].

TEs are traditionally classified according to their replication mode [7]. Elements using an RNA molecule as an intermediate are called Class I or retrotransposons, while elements using a DNA intermediate are called Class 2 or transposons [8]. Each class of TEs is further sub-classified by a hierarchical system into orders, superfamilies, lineages, and families [9].

Several bioinformatic methods were developed to detect TEs in genome sequences, including homology-based, de novo, structure-based, and comparative genomic, but no combination of them can provide a reliable detection in a relatively short time [10]. Most of the algorithms currently available use a homology-based approach [11], displaying performance issues when analyzing elements in large plant genomes. In the current scenario of large-scale sequencing initiatives, such as the Earth BioGenome Project [12], disruptive technologies and innovative algorithms will be necessary for genome analysis in general and, particularly, for the detection and classification of TEs that represent the main portion of these genomes [13].

In recent years, several databases consisting of thousands of TE at all classification levels of several species and taxa have been created and published [3]. Furthermore, these databases have different characteristics, such as containing consensus [14–16] or genomic [17,18] TE sequences, coding domains [9,19], and also TE-related RNA [20,21]. These databases have been constructed with the TEs detected in species sequenced using bioinformatics approaches (commonly based on homology or structure), which can produce false positive if there is no a curation process [11]. As other biological sets (such as datasets of splice sites [22], or protein function predictions [23]), databases have distinct numbers of different types of TEs producing unbalanced classes [23]. For example in PGSB, the largest proportion of the elements corresponds to retrotransposons (at least 86%) [24]. The above is caused by the replication mode of each TE class. As in other detection tasks, the negative instances for identifying TEs are all other genomic elements than TEs (that constitute the positive instances) [25–27], such as introns, exons, CDS (coding sequences), and simple repeats, among others, making the negative class multimodal. These databases constitute valuable resources to improve tasks like TE detection and classification using bioinformatics or also novel techniques such as machine learning (ML).

ML is defined as a set of algorithms that can be calibrated based on previously processed data or past experience [28] and a loss function through an optimization process [29] to build a model. ML is applied to different bioinformatics problems, including genomics [30], systems biology, evolution [28], and metagenomics [31], demonstrating substantial benefits in terms of precision and speed. Several recent studies using ML to detect TEs report drastic improvements in the results [32–34] compared to conventional bioinformatics algorithms [13].

In ML, the selection of adequate metrics that measure the algorithms' performance is one of the most crucial and challenging steps. Commonly used metric for classification tasks are accuracy, precision, recall, and ROC curves [35,36], but they are not appropriate for all datasets [37], especially when the positive and negative datasets are unbalanced [13]. Accuracy and ROC curves can be meaningless performance measurements in unbalanced datasets [22], because it does not reveal the true classification performance of the rare classes [38]. For example, ROC curves are not commonly used in TE classification, because only a small portion of the genome contains certain TE superfamilies [34]. On the other hand, precision and recall can be more informative since precision is the percentage of predictions that are correct [34] and recall is the percentage of true samples that are correctly detected [26], nevertheless it is recommended to use them in combination with other metrics since the use of only one of these metrics cannot provide a full picture of the algorithm performance [36].

Most of the classification and detection tasks addressed by ML define two classes, positive and negative [13]. Thus, expected results can be classified as true positive (tp) if they were classified as positive and are contained in the positive class, while as false negatives (fn) if they were rejected but did not belong to the negative class. On the other hand, samples that are contained in negative class and predicted to be positive constitute false positives (fp), or true negative (tn) if they are not [13,28,39]. These markers are related in the confusion matrix, and most of the metrics used in ML are calculated based on this matrix.

Depending on the goal of the application and the characteristics of the elements to be classified, other metrics addressing classification (binary, multiclass, hierarchical), class balance (i.e., if training dataset is imbalanced or not), and the importance of positive or negative instances [36] must be considered. Another point is the ability of a metric to preserve the value under a change in the confusion matrix, called measure invariance [40]. This properties give comparative parameters between metrics that are not based on datasets, but in the way they are calculated. Each of the properties of the invariance can be beneficial or unfavorable depending on the main objectives, the balance of the classes, the size of the data sets, the quality, and the composition of the negative class, among others [40]. Thus, invariance properties are useful tools in order to select the most informative metrics in each ML problem.

Recently, different ML-based software have been developed to tentatively detect repetitive sequences [34,41,42], classify them (at the order or superfamily levels) [27,43–45], or both [10,46]. Additionally, deep neural networks-based software were also developed to classify TEs [11,47]. Nevertheless, there are no studies about which metrics can be more suitable taking into account the unique characteristics of transposable element datasets and their dynamic structure. Here, we evaluated 26 metrics found in the literature for TE detection and classification, considering the main features of this type of data, the invariance properties and characteristics of each metric in order to select the more appropriate ones for each type of classification.

2. Materials and Methods

2.1. Bibliography Analysis

As a literature information source, we used the results obtained by [13], who applied the systematic literature review (SLR) process proposed by [48]. The authors applied the search Equation (1) to perform a systematic review of research articles, book chapters and other review papers presented in well-known bibliographic databases such as Scopus, Science Direct, Web of Science, Springer Link, PubMed, and Nature.

$$\text{(“transposable element” OR retrotransposon OR transposon) AND (“machine learning” OR “deep learning”)} \tag{1}$$

Applying the Equation (1), a total of 403 publications were identified of which authors removed those which do not satisfy certain conditions such as repeated (the same study was found in different databases); of different types (books, posters, short articles, letters and abstracts); and written in other languages (languages other than English). Then, authors used inclusion and exclusion criteria in order to select interested articles. Finally, 35 publications were selected as relevant in the fields of ML and TE [13]. Using these relevant publications, we identified the metrics used for the detection and classification of TEs, preserving information such as representation and observations (i.e., the properties measured). Next, we evaluated each metric that was reported as a decisive source in relevant publications. The characteristics and properties of each metric were analyzed regarding their application to TEs, considering that these elements have some characteristics, such as highly variant dynamics for each class, negative datasets with a large number of genomic elements for detection, a great divergence between elements of the same class, and species specificity.

2.2. Measure Invariance Analysis

Comparing the performance measures in ML approaches is not straightforward, and although the most common way to select most informative measures is by using empirical analysis [49,50], an alternative methodology was proposed [40], which consists of assessing whether a given metric changes its value under certain modifications in the confusion matrix. This property is named measure invariance, and can be used to compare performance metrics without focusing on their experimental results but using their measuring characteristics such as detecting variations in the number of true positives (tp), false positives (fp), false negative (fn), or true negatives (tn) presented in the confusion matrix [40]. Thus, a measure is invariant when its calculation function f which receives a confusion matrix produces the same value even if the confusion matrix has modifications. For example, consider the following confusion matrix $m = \begin{bmatrix} 10 & 4 \\ 3 & 16 \end{bmatrix}$, where tp = 10, fn = 4, fp = 3, and tn = 16 and the function for calculating accuracy $f = \frac{tp+tn}{tp+fp+fn+tn}$, thus the accuracy for the confusion matrix presented above is $f(m) = 0.78$. Now consider exchanging the positive (tp by tn) and negative (fp by fn) values in the confusion matrix obtaining the following $m' = \begin{bmatrix} 16 & 3 \\ 4 & 10 \end{bmatrix}$. If we apply the function f over the new confusion matrix, so we obtain $f(m') = 0.78$. In this case, we can conclude that accuracy cannot detect exchanges of positive and negative values and thus it is invariant due to $f(m) = f(m')$.

In this work, we used eight invariance properties to compare measures which were selected in the bibliographic analysis. All these invariances were derived from basic matrix operations, such as addition, scalar multiplication, and transposition of rows or columns, as following [40]:

- Exchange of positives and negatives (I1): A measure presents invariance in this property if $f\left(\begin{bmatrix} tp & fn \\ fp & tn \end{bmatrix}\right) = f\left(\begin{bmatrix} tn & fp \\ fn & tp \end{bmatrix}\right)$, showing invariance corresponding to the distribution of classification results due to its inability to differentiate tp from tn and fn from fp. An invariant metric in this property may not be utilized in datasets highly unbalanced [40], such as the number of TEs belonging to each lineage in the Repbase or PGSB databases.

- Change of true negative counts (I2): A measure presents invariance in this property if $f\left(\begin{bmatrix} tp & fn \\ fp & tn \end{bmatrix}\right) = f\left(\begin{bmatrix} tp & fn \\ fp & tn' \end{bmatrix}\right)$, demonstrating the inability to recognize specificity of the classifiers. This property can be useful in problems with multi-modal negative class (the class with all elements other than the positive), i.e., in the detection of TEs, where negative class may be composed by all other genomic features such as genes, CDS (coding sequences), and simple repeats, among others.

- Change of true positive counts (I3): A measure presents invariance in this property if $f\left(\begin{bmatrix} tp & fn \\ fp & tn \end{bmatrix}\right) = f\left(\begin{bmatrix} tp' & fn \\ fp & tn \end{bmatrix}\right)$, losing the sensitivity of the classifiers, so their evaluation should be complementary to other metrics.

- Change of false negative counts (I4): A measure presents invariance in this property if $f\left(\begin{bmatrix} tp & fn \\ fp & tn \end{bmatrix}\right) = f\left(\begin{bmatrix} tp & fn' \\ fp & tn \end{bmatrix}\right)$, indicating stability even when the classifier has errors assigning negative labels. It is helpful in detecting or classifying TEs when non-curated databases are used in training (such as RepetDB), which may contain mistakes.

- Change of false positive counts (I5): A measure presents invariance in this property if $f\left(\begin{bmatrix} tp & fn \\ fp & tn \end{bmatrix}\right) = f\left(\begin{bmatrix} tp & fn \\ fp' & tn \end{bmatrix}\right)$, proving reliable results even though some classes contain outliers, which is common in elements classified at lineage level due to TE diversity in their nucleotide sequences [26].

- Uniform change of positives and negatives (I6): A measure presents invariance in this property if $f\left(\begin{bmatrix} tp & fn \\ fp & tn \end{bmatrix}\right) = f\left(\begin{bmatrix} k1tp & k1fn \\ k1fp & k1tn \end{bmatrix}\right)$, with $k1 \neq 1$. It indicates if a measure's value changes when the size of the dataset increases. The non-invariance indicates that the application of the metric depends on size of the data.

- Change of positive and negative columns (I7): A measure presents invariance in this property if $f\left(\begin{bmatrix} tp & fn \\ fp & tn \end{bmatrix}\right) = f\left(\begin{bmatrix} k1tp & k1fn \\ k2fp & k2tn \end{bmatrix}\right)$, with $k1 \neq k2$. If a metric is unchanged in this way, it will not show changes when additional datasets differs from training datasets in quality (i.e., having more noise), and indicating the needed of other measures as complement. On the contrary, if a metric presents a non-invariant behavior then, it may be suitable if different performances are expected across classes.

- Change of positive and negative rows (I8): A measure presents invariance in this property if $f\left(\begin{bmatrix} tp & fn \\ fp & tn \end{bmatrix}\right) = f\left(\begin{bmatrix} k1tp & k1fn \\ k2fp & k2tn \end{bmatrix}\right)$, with $k1 \neq k2$. In this case, if a metric is non-invariant, its applicability depends on the quality of the classes. It may be useful, for example, when curated datasets are available such as Repbase.

Properties described above were calculated by [40] for commonly used performance measures and we used them to analyze selected metrics (Table 1), except for area under the precision-recall curve (auPRC) which was calculated by us, following the methodology proposed by authors.

Table 1. Invariance properties of selected metrics. 0 for invariance and 1 for non-invariance. Adapted from [40].

Metric	I1	I2	I3	I4	I5	I6	I7	I8
F1-score	0	1	0	0	0	1	0	0
auPRC *	1	0	0	0	0	1	0	1
Fscoreμ	0	1	0	0	0	1	0	0
PrecisionM	0	1	0	1	0	1	1	0
RecallM	0	1	0	0	1	1	0	1
FscoreM	0	1	0	0	0	1	0	0
Precision↓	0	1	0	1	0	1	1	0
Recall↓	0	1	0	0	1	1	0	1
Fscore↓	0	1	0	0	0	1	0	0
Fscore↑	0	1	0	0	0	1	0	0

* The invariance properties of this metric were calculated by authors in this study. I1: Exchange of positives and negatives, I2: Change of true negative counts, I3: Change of true positive counts, I4: Change of false negative counts, I5: Change of false positive counts, I6: Uniform change of positives and negatives, I7: Change of positive and negative columns, and I8: Change of positive and negative rows.

2.3. Experimental Analysis

To test the behavior of the most commonly used metrics, such as accuracy, precision, and recall, and the best scoring metric found in this study, we performed several experiments addressing the specific problem of multi-class classification of LTR retrotransposons at the lineage level in plants. We selected this problem since LTR retrotransposons are the most common repeat sequences in almost all angiosperms and they represent an important fraction of their host genome; for instance, 75% in maize [51], 67% in wheat [52], 55% in *Sorghum bicolor* [53], and 42% in Robusta coffee [54]. As input, we used two well-known TE databases: Repbase (free version, 2017) [14] and PGSB [17]. For Repbase, we joined the LTR domains with the internal section (concatenating before and after) of each LTR retrotransposon found in the database. The first step was to generate a well-curated dataset of LTR retrotransposons; thus, we classified LTR retrotransposons from both databases at the lineage level using the homology-based Inpactor software [55] with RexDB nomenclature [9]. Inpactor has two filters for deleting nested elements: (1) Removing elements with domains belonging to two different superfamilies (i.e., Copia and Gypsy) and (2) removing elements with domains belonging to two or more different lineages. Additionally, we applied three extra filters: (1) Removing elements with lengths different from those reported by the Gypsy Database [19] with a tolerance of 20% (this value was chosen to filter elements with nested insertion of others TEs but keeping elements with natural divergence), (2) removing elements with less than two domains (incomplete elements derived from deletion processes), and (3) removing elements with insertions of partial or complete TEs from class II (present in Repbase). Finally, we removed elements from the following lineages: Alesia, Bryco, Lyco, Gymco, Osser, Tar, CHLAMYVIR, Retand, Phygy, and Selgy due to their very low frequency or absence in angiosperms.

Since the datasets used in this study are categorical (nucleotide sequences), we transformed them using the coding schemes shown in Table 2. Also, we used two additional techniques to automatically extract features from the sequences; (1) for each element, we obtained k-mer frequencies using k values between one and six (this range of values of k was selected due to k-mers with k > 6 are rare in sequences and probably do not provide informational features and they are computationally expensive to calculate) and (2) we extracted three physical-chemical (PC) properties, such as average hydrogen bonding energy per base pair (bp), stacking energy (per bp), and solvation energy (per bp), which are calculated by taking the first di-nucleotide and then moving in a sliding window of one base at a time [56]. Since the ML algorithms used here require sequences of the same lengths, we found the largest TE in each dataset and completed the smaller sequences by replicating their nucleotides.

Table 2. Coding schemes for translating DNA characters in numerical representations. Adapted from [13].

Coding Scheme	Codebook	Reference
DAX	{'C':0, 'T':1, 'A':2, 'G':3}	[57]
EIIP	{'C':0.1340, 'T':0.1335, 'A':0.1260, 'G':0.0806} {'C':−1, 'T':−2, 'A':2, 'G':1}	[58]
Complementary	{'C':−1, 'T':−2, 'A':2, 'G':1}	[59]
Enthalpy	{'CC':0.11, 'TT':0.091, 'AA':0.091, 'GG':0.11, 'CT':0.078, 'TA':0.06, 'AG':0.078, 'CA':0.058, 'TG':0.058, 'CG':0.119, 'TC':0.056, 'AT':0.086, 'GA':0.056, 'AC':0.065, 'GT':0.065, 'GC':0.1111}	[60]
Galois (4)	{'CC':0.0, 'CT':1.0, 'CA':2.0, 'CG':3.0, 'TC':4.0, 'TT':5.0, 'TA':6.0, 'TG':7.0, 'AC':8.0, 'AT':9.0, 'AA':1.0, 'AG':11.0, 'GC':12.0, 'GT':13.0, 'GA':14.0, 'GG':15.0}	[61]

We applied the workflow described in [62] to compare commonly used ML algorithms using supervised techniques. As the authors suggested, we applied four types of pre-processing strategies: none (raw data), scaling, data dimensionality reduction using principal component analysis (PCA), and both scaling and PCA. On the other hand, we used some of the most common ML algorithms [62], including linear support vector classifier (SVC), logistic regression (LR), linear discriminant analysis (LDA), K-nearest neighbors (KNN), naive Bayesian classifier (NB), multi-layer perceptron (MLP), decision trees (DT), and random forest (RF). All algorithms were tested by varying or tuning parameter values to find the best performance (Table 3).

The experiments consisted in executing all possible combinations between databases, coding schemes, pre-processing strategies, and ML algorithms (Figure 1 and Table 4). First, we used the accuracy and, the F1-score using the macro-averaging strategy as main metric in tuning process (Table 3). Finally, we calculated other common metrics using the best value of the tuned parameter in each algorithm for comparison. All the experiments were performed using Python 3.6 and Scikit-Learn library 0.22 [63], installed in a Anaconda environment in Linux over a CPU architecture. We ran our tests using the HPC cluster of IFB (https://www.france-bioinformatique.fr), IRD itrop (https://bioinfo.ird.fr/) and Genotoul Bioinformatics platform (http://bioinfo.genotoul.fr/), all of them are managed by Slurm.

Table 3. Tested algorithm parameters.

Algorithm	Parameter	Range	Step	Description
KNN	n_neighbors	1–99	1	Number of neighbors
SVC	C, gamma = 1×10^{-6}	10–100	10	Penalty parameter C of the error term.
LG	C	0.1–1	0.1	Inverse of regularization strength
LDA	tol	0.0001–0.001	0.0001	Threshold used for rank estimation in SVD solver.
NB	var_smoothing	1×10^{-1}–1×10^{-19}	1×10^{-2}	Portion of the largest variance of all features that is added to variances for calculation stability.
MLP	Solver = 'lbfgs', alpha = 0.5, hidden_layer_sizes	50–1050	50	Number of neurons in hidden layers. In this study, we used solver lbfgs and alpha 0.5
RF	n_estimators	10–100	10	The number of trees in the forest.
DT	max_depth	1–10	1	The maximum depth of the tree.

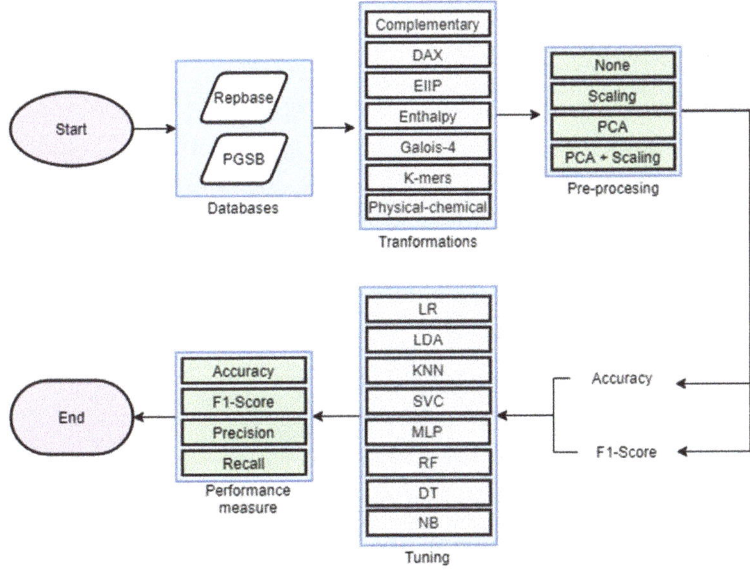

Figure 1. Overall flow of the experimental analysis done in this work.

Table 4. Description of experiments performed.

Experiment ID	Database	Algorithm	Pre-Processing	Main Metric
Exp1	Repbase	LR, LDA, MLP, KNN, DT, RF, SVM, NB	None, Scaling, PCA, Scaling + PCA	Accuracy
Exp2	Repbase	LR, LDA, MLP, KNN, DT, RF, SVM, NB	None, Scaling, PCA, Scaling + PCA	F1-score
Exp3	PGSB	LR, LDA, MLP, KNN, DT, RF, SVM, NB	None Scaling, PCA, Scaling + PCA	Accuracy
Exp4	PGSB	LR, LDA, MLP, KNN, DT, RF, SVM, NB	None, Scaling, PCA, Scaling + PCA	F1-score

3. Results

3.1. Bibliography and Invariance Analysis

Based on relevant literature sources (articles) detected in [13], by searching in several databases, we collected 26 metrics that are commonly used in different types of classification tasks, such as binary, multi-class, and hierarchical (Table 5). We were interested in classification metrics because the detection task can be considered as a binary classification (using TEs as positive class and non-TEs as negative class). Additionally, we assigned an importance level for each metric (Table 5) based on the following aspects: (i) How appropriate is its application to analyzing TE datasets (detection and classification)? (ii) Which features are measured and how important are these features for TE analysis? For each metric, each aspect is assigned a level of importance (low, medium, high). Furthermore, the properties reported in relevant publications were used to evaluate each metric. In this way, we extracted and summarized information about each metric and we evaluated if its use for TE datasets is plausible. General observations of metrics can be found in the observations column in Table S4.

Table 5. Metrics used in classification problems. Adopted from [22,34,35,40,64–68].

ID	Metric	Classification Type	Used in TEs	Level of Applicability to TEs	Level of Mmeasured Features
1	Accuracy	Binary	[10,32,45,69,70]	Low	Low
2	Precision (Positive predictive value)	Binary	[34]	Medium	Medium
3	Sensitivity (recall or true positive rate)	Binary	[10,32,34,71]	Medium	Medium
4	Specificity	Binary	[71]	Low	Low
5	Matthews correlation coefficient	Binary	NO	High	Low
6	Performance coefficient	Binary	NO	Low	Low
7	**F1-score**	**Binary**	**[34,47,72]**	**High**	**High**
8	Precision-recall curves	Binary	[25,34]	High	High
9	Receiver Operating Characteristic curves (ROCs)	Binary	[71]	Low	Low
10	Area under the ROC curve (AUC) [a]	Binary	[25,70]	Low	Low
11	**Area under the Precision Recall Curve (auPRC)[b]**	**Binary**	**NO**	**High**	**High**
12	False-positive rate	Binary	[70,71]	Medium	Low
13	Average Accuracy	Multiclass	[42]	Low	Low
14	Error Rate	Multiclass	NO	Low	Low
15	Precisionμ	Multiclass	NO	Medium	Low
16	Recallμ	Multiclass	NO	Medium	Low
17	**Fscoreμ**	**Multiclass**	**NO**	**High**	**Low**
18	**PrecisionM**	**Multiclass**	**[34,43]**	**Medium**	**Medium**
19	**RecallM**	**Multiclass**	**NO**	**Medium**	**Medium**
20	**FscoreM**	**Multiclass**	**NO**	**High**	**High**
21	**Precision$\downarrow$**	**hierarchical**	**[11,23,24]**	**Medium**	**Low**
22	**Recall$\downarrow$**	**hierarchical**	**[11,23,24]**	**Medium**	**Low**
23	**Fscore$\downarrow$**	**hierarchical**	**[11,23,24]**	**High**	**Low**
24	Precision$\uparrow$	hierarchical	[11,23,24,27]	Medium	Medium
25	Recall$\uparrow$	hierarchical	[11,23,24,27]	Medium	Medium
26	**Fscore$\uparrow$**	**hierarchical**	**[11,23,24,27]**	**High**	**High**

Although [a] and [b] are areas under the curve, they can be viewed as a linear transformation of the Youden Index [73]. Rows in bold were selected to perform invariance analyses. Additional information, such as metric representation and general observations of this table, is available in Table S1–S3 and observations about levels of applicability and measured features can be found in Table S4: Rows in bold were selected to perform invariance analyses.

We performed invariance analyses on the metrics with the best evaluation for each classification type (Table 1). Precision-Recall curves were excluded for further analysis at this step since it is impossible to calculate graphics from a confusion matrix. We obtained the invariance properties for almost all metrics from [40], except for area under the precision recall curve (ID = 11). For this metric, we generated a random confusion matrix and applied all the transformations presented in [40] in order to calculate its value and determine if it changed or not. The invariance analyses were performed based on the one described by [40].

3.2. Experimental Analysis

To evaluate the relevance of the literature reports about metrics, we applied them to experiments on the multi-class classification of LTR retrotransposons in angiosperm plants at the lineage level. Using nucleotide sequences from Repbase [14] (free version, 2017) and PGSB [17] as input, we performed a classification process using Inpactor [55]. We generated high-quality datasets by removing sequences that did not satisfy certain filters (See Materials and Methods). After filtering and homology-based classification, we obtained 2,842 TEs from Repbase and 26,371 elements from PGSB (Table 6).

Table 6. Number of nucleotide sequences of each plant lineage used in Repbase and PGSB databases.

Lineage	Repbase	PGSB
ALE	53	230
ANGELA	32	1344
ATHILA	107	1844
BIANCA	36	319
CRM	101	1041
DEL	162	2738
GALADRIEL	27	109
IKEROS	0	59
IVANA	7	7
ORYCO	438	1169
REINA	551	1086
RETROFIT	781	1151
SIRE	63	4393
TAT	203	9578
TEKAY	0	11
TORK	281	1292
TOTAL	2842	26,371

We executed four experiments using the generated datasets (Table 4) to evaluate the behavior of each metric in different configurations. In the first two experiments, we were interested in analyzing the performance of accuracy and F1-score metrics using a well-curated dataset (Repbase) but with a few different sequences in some lineages (Figure 2, Figures S1 and S2). In the last two experiments, we evaluated a larger dataset (PGSB) and tested the same two metrics (Figure 3, Figure S3 and S4). The complete results of all the experiments can be consulted in Tables S5–S8.

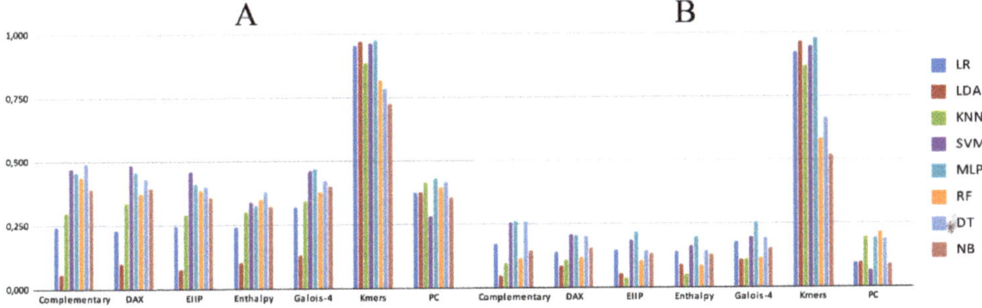

Figure 2. Performance of machine learning (ML) algorithms and Repbase pre-processed data by principal component analysis (PCA) and scaling processes using as main metric: (**A**) accuracy and (**B**) F1-score.

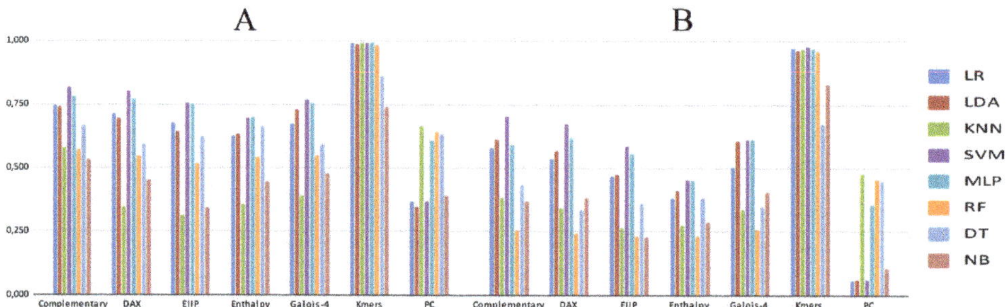

Figure 3. Performance of ML algorithms and PGSB pre-processed data by PCA and scaling processes using as main metric: (**A**) Accuracy and (**B**) F1-score.

Figures 2 and 3 show the best performance achieved by each algorithm after tuning one parameter (Table 3), using as main metric accuracy or F1-score. Since each coding scheme displayed a different behavior, we were interested in further analyzing how each metric behaves in different algorithms and coding schemes. K-mers (Figure 4) showed the best performance, PC (Figure 5) displayed the worst performances, and complementary (Figure 6) showed an intermediate performance, which were selected for further analyses.

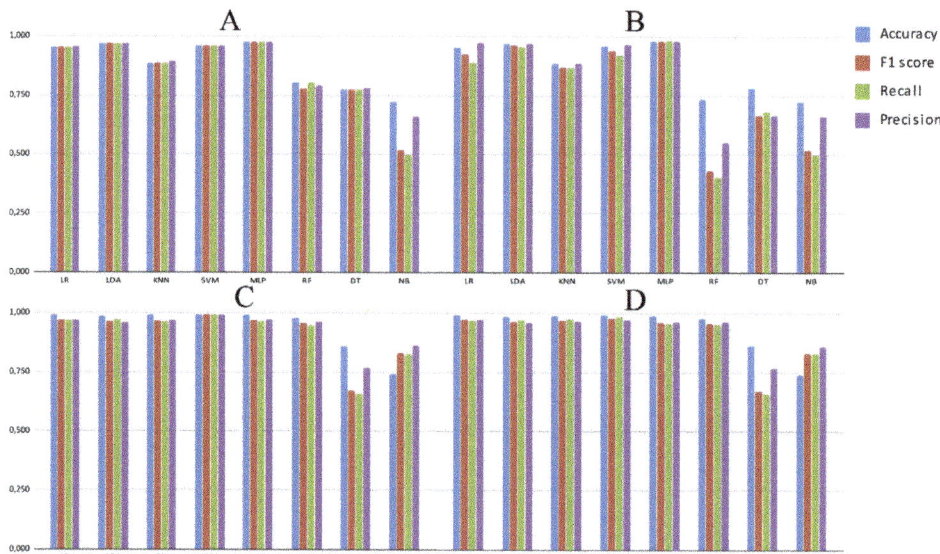

Figure 4. Results of ML algorithms using nucleotide sequences transformed by k-mers, PCA and scaling, and applying accuracy, F1-score, recall and precision metrics. Experiments: (**A**) Exp1, (**B**) Exp2, (**C**) Exp3, and (**D**) Exp4.

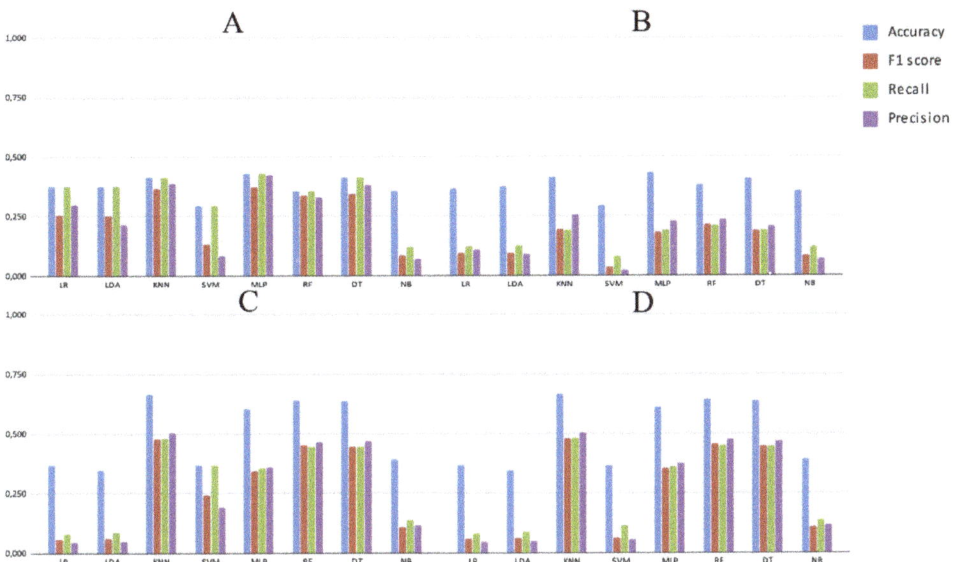

Figure 5. Results of ML algorithms using sequences transformed by PC and PCA and scaling, and applying accuracy, F1-score, recall, and precision metrics. Experiments: (**A**) Exp1, (**B**) Exp2, (**C**) Exp3, and (**D**) Exp4.

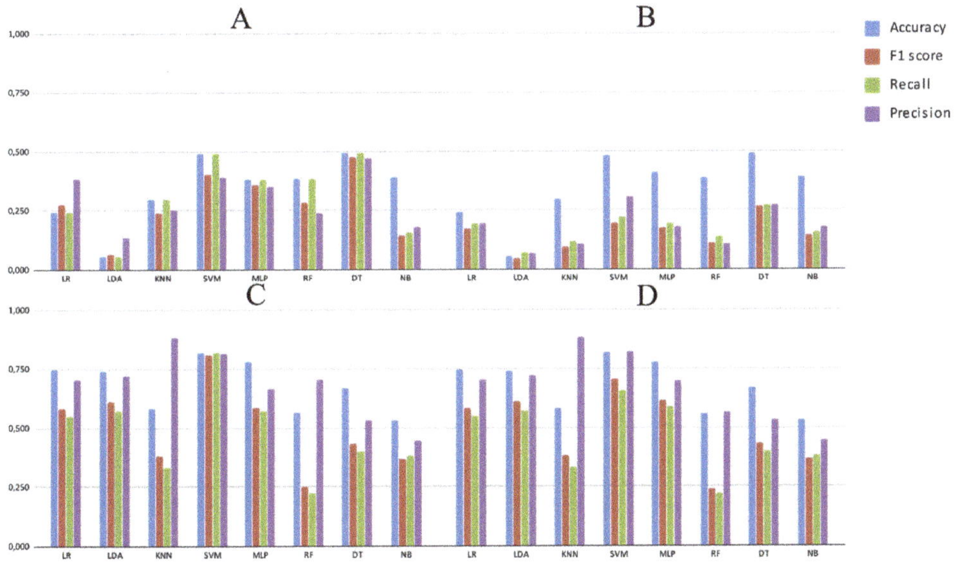

Figure 6. Results of ML algorithms using sequences transformed by PC and PCA and scaling, and applying accuracy, F1-score, recall and precision metrics. Experiments: (**A**) Exp1, (**B**) Exp2, (**C**) Exp3, and (**D**) Exp4.

4. Discussion

The detection and classification of transposable elements is a crucial step in the annotation of sequenced genomes, because of their relation with genome evolution, gene function, regulation, and alteration of expression, among others [74,75]. This step remains challenging given their abundance and diverse classes and orders. In addition, other characteristics of TEs, such as a relatively low selection pressure and a more rapid evolution than coding genes [26], their dynamic evolution due to insertions of other TEs (nested insertion), illegitimate and unequal recombination, cellular gene capture, and inter-chromosomal and tandem duplications [76], make them difficult targets for accurate and rapid detection and classification procedures. Indeed, TEs showing uniform structures and well-established mechanisms of transposition can be easily clustered and classified into major groups such as orders or superfamilies (e.g., LTR retrotransposons) [77]. However, this task is relatively complex and time-consuming when classifying TEs into lower levels, such as lineages or families [78]. For these reasons, TE classification and annotation are complex bioinformatics tasks [79], in which, in some cases, manual curation of sequences is required by specialists. The ability of biologists to sequence any organism or a group of organisms in a relatively short time and at relatively low costs redefines the barrier of the genomic information. The current limitation is not the generation of genome sequences but the amount of information to be processed in a limited time. Complex bioinformatics tasks may be accomplished by machine learning algorithms, such as in drug discovery and other medical applications [80], genomic research [38,81], metagenomics [31,82], and multiple applications in proteomics [83].

Previous works apply ML and DL for TE analysis, such as Arango-López et al. (2017) [43] for the classification of LTR-retrotransposons, Loureiro et al. (2012) [84] for the detection and classification of TEs using developed bioinformatics tools, and Ashlock and Datta (2012) [69] distinguishing between retroviral LTRs and SINEs (short interspersed nuclear elements). Deep neural networks (DNN) are also used to hierarchically classify TEs by applying fully connected DNN [11] and through convolutional neural networks (CNN) and multi-class approaches [47].

In TE detection and classification, the dataset could be highly imbalanced [23]; therefore, commonly used metrics such as accuracy and ROC curves may not be fully adequate [36]. For the detection task, the positive class will be much lower than the negative, because the latter will have all other genomic elements. In classification, each type of TE (classes, orders, superfamilies, lineages, or families) has different dynamics that produce a distinct number of copies. For example, in the coffee genus, LTR-retrotransposons show large copy number differences depending on the lineage [85]. In *Oryza australiensis* [86] and pineapple genomes [87], only one family of LTR-retrotransposons contributes to 26% and 15% (Pusofa) of the total genome size, respectively.

For binary classification (for example, to detect TEs or classify them into class 1 and class 2), the most appropriate metric is F1-score (id = 7), which considers precision and recall values. Precision is a useful parameter when the number of false-positive must be limited and recall measures how many positive samples are captured by the positive predicted [36]. However, the use of only one of these metrics cannot provide a full picture of the algorithm performance. Altogether, our results suggest that F1-score is appropriate for TE analyses.

In multi-class approaches (such as TE classification into orders, superfamilies, or lineages), F1-score (id = 20) also seems to be the most suitable metric, combined with the macro-averaging strategy, probably due to the high diversity of intra-class samples. For TE detection and classification, it appears more important to weigh all classes equally than to weigh each sample equally (micro-averaging strategy). Finally, for hierarchical classification approaches (i.e., considering the hierarchical classification of TEs proposed by Wicker and coworkers [8]), F1-score↓ (id = 26) and F1-score↑ (id = 23) seem most suitable. These results demonstrate the importance of calculating the performance of each hierarchical level. Additionally, precision-recall curves and area under the precision-recall curve provided the best results for binary classification, demonstrating that, for TE datasets, they are more appropriate than the commonly used ROC curves.

Area under the precision-recall curve, auPRC (id = 11), is a unique metric, which showed invariance in I1 and non-invariance in I2. Its invariance properties make auPRC a robust measure of the overall performance of an algorithm and it is insensitive to the performance for a specific class (I1). However, it less appropriate for data with a multi-modal negative class (~I2).

All metrics presented invariance in I3, indicating that they could not measure true positive change. This suggests that they can be used when the positive class is not very strong. PrecisionM (id = 18) and Precision↓(id = 21) showed non-invariance in I4, which demonstrates that these metrics may be less reliable when manual labeling follows rigorous rules for a negative class. On the other hand, RecallM (id = 19) and Recall↓ (id = 22) exhibited non-invariance in I5, indicating that these metrics may not provide a conservative estimate when the positive class has outliers, as commonly found in TE datasets. Thus, these metrics might not be informative in TE detection and classification. The non-invariance properties of all metrics in I6, shown in Table 1, demonstrated that these metrics can vary in data with large size differences. Consequently, these metrics must be used carefully for comparison with other and different datasets.

Non-invariance in I7 shown by precision (id = 18 and 21) supported the combined use of this metric with other metrics (such as in F1-score) common in ML algorithms. Finally, auPRC (id = 11), RecallM (id = 19), and Recall↓ (id = 22) may be better choices for the evaluation of classifiers if different data sizes exhibit the same quality of positive (negative) characteristics, as in the case of generated (simulated) data due to their non-invariance properties in I8.

Our tests for the multi-class classification task of LTR retrotransposons at the lineage level show an overestimation of the performance of all ML algorithms used here (Figures 2 and 3) for both datasets (Repbase and PGSB). Furthermore, our experiments support the information found in the literature, indicating that accuracy is not the most informative metric for highly unbalanced datasets, such as those used in this study. Additionally, Figures 2 and 3 indicate that this tendency of overestimation is generalized for nearly all the algorithms, pre-processing techniques, and coding schemes used here.

A clear exception, however, is shown by k-mers (in both training and validation datasets, Tables S4–S7), for which accuracy and F1-scores did not show any differences. Nevertheless, if the F1-score is used in the tuning process (Figure 4B,D, Figure 5B,D, and Figure 6B,D), accuracy also overestimates the performance of almost all the algorithms in comparison to F1-score, sensitivity (recall), and precision. Interestingly, RF performs in a similar manner to that of the other algorithms when PGSB (with more than 26,000 elements) is used, but DT presents the same behavior in both datasets.

When the performance of a given scheme is low, the overestimation shown by accuracy is more evident (Figures 5 and 6). This is due to the extremely low performance on some lineages and, thus, accuracy is not very informative if it is not used combined with another metric. As suggested by the literature and invariance analyses, F1-score appears to be the most adequate and informative metric in the experiments performed here, since it is a harmonic estimate of precision and sensitivity by measuring the combined amount of false-positive and positive samples captured by the algorithm.

Overall, the results shown here can also be applied to data similar to TEs, such as retrovirus and endogenous retrovirus or data with highly imbalanced classes, high intra-class diversity, and negative multi-modal classes (in detection tasks).

5. Conclusions

Altogether, our analyses suggest that F1-score may be the best metric for each of the ML classification types, except for simulated data, for which auPRC and Recall should be more appropriate because of their invariance properties. Conversely, precision should be used in combination with other metrics to avoid non-realistic estimates of algorithm performance. In binary classification, precision-recall curves must be used instead of ROC curves. In multi-class classification approaches, the macro-averaging strategy seems to be more appropriate for TE detection and classification. As future work, we propose to develop a ML model based on the databases, algorithms, and coding schemes used here and using F1-score in the tuning process, to improve classification of LTR retrotransposons at the lineage level in angiosperms.

Supplementary Materials: The following are available online at http://www.mdpi.com/2227-9717/8/6/638/s1: Figure S1. Performance of ML algorithms and Repbase using accuracy as the main metric (experiment 1) and the following pre-processing techniques: (a) None, (b) scaling, (c) PCA, (d) PCA + scaling. Figure S2. Performance of

ML algorithms and Repbase using F1-score as the main metric (experiment 2) and the following pre-processing techniques: (a) None, (b) scaling, (c) PCA, (d) PCA + scaling. Figure S3. Performance of ML algorithms and PGSB using accuracy as the main metric (experiment 3) and the following pre-processing techniques: (a) None, (b) scaling, (c) PCA, (d) PCA + scaling. Figure S4. Performance of ML algorithms and PGSB using F1-score as the main metric (experiment 4) and the following pre-processing techniques: (a) None, (b) scaling, (c) PCA, (d) PCA + scaling. Table S1. Metrics used in binary classification. Adopted from [22,34,35,40,64–68]. Table S2. Metrics used in multi-class classification. Adopted from [22,34,35,40,64–68]. Table S3. Metrics used in hierarchical classification. Adopted from [22,34,35,40,64–68]. Table S4. Evaluation for metric collection. Table S5. Results of experiment 1. Table S6. Results of experiment 2. Table S7. Results of experiment 3. Table S8. Results of experiment 4.

Author Contributions: Conceptualization, S.O.-A., G.I., and R.G.; methodology, S.O.-A., J.S.P., and R.T.-S.; writing—original draft preparation, S.O.-A., R.T.-S., J.S.P., L.F.C.-O., G.I., and R.G.; writing—review and editing, S.O.-A., R.T.-S., J.S.P., L.F.C.-O., G.I., and R.G.; supervision, G.I. and R.G. All authors have read and agreed to the published version of the manuscript.

Funding: Simon Orozco-Arias is supported by a Ph.D. grant from the Ministry of Science, Technology and Innovation (Minciencias) of Colombia, Grant Call 785/2017. The authors and publication fees were supported by Universidad Autónoma de Manizales, Manizales, Colombia under project 589-089, and Romain Guyot was supported by the LMI BIO-INCA. The funders had no role in the study design, data collection and analysis, the decision to publish, or preparation of the manuscript.

Acknowledgments: The authors acknowledge the IFB Core Cluster that is part of the National Network of Compute Resources (NNCR) of the Institut Français de Bioinformatique (https://www.france-bioinformatique.fr), the Genotoul Bioinformatics platform (http://bioinfo.genotoul.fr/), and the IRD itrop (https://bioinfo.ird.fr/) at IRD Montpellier for providing HPC resources that have contributed to the research results reported in this paper.

Conflicts of Interest: The authors declare no conflict of interest.

References

1. Mita, P.; Boeke, J.D. How retrotransposons shape genome regulation. *Curr. Opin. Genet. Dev.* **2016**, *37*, 90–100. [CrossRef] [PubMed]

2. Keidar, D.; Doron, C.; Kashkush, K. Genome-wide analysis of a recently active retrotransposon, Au SINE, in wheat: Content, distribution within subgenomes and chromosomes, and gene associations. *Plant Cell Rep.* **2018**, *37*, 193–208. [CrossRef] [PubMed]

3. Orozco-Arias, S.; Isaza, G.; Guyot, R. Retrotransposons in Plant Genomes: Structure, Identification, and Classification through Bioinformatics and Machine Learning. *Int. J. Mol. Sci.* **2019**, *20*, 781. [CrossRef] [PubMed]

4. De Castro Nunes, R.; Orozco-Arias, S.; Crouzillat, D.; Mueller, L.A.; Strickler, S.R.; Descombes, P.; Fournier, C.; Moine, D.; de Kochko, A.; Yuyama, P.M.; et al. Structure and Distribution of Centromeric Retrotransposons at Diploid and Allotetraploid Coffea Centromeric and Pericentromeric Regions. *Front. Plant Sci.* **2018**, *9*, 175. [CrossRef] [PubMed]

5. Ou, S.; Chen, J.; Jiang, N. Assessing genome assembly quality using the LTR Assembly Index (LAI). *Nucleic Acids Res.* **2018**, 1–11. [CrossRef] [PubMed]

6. Mustafin, R.N.; Khusnutdinova, E.K. The Role of Transposons in Epigenetic Regulation of Ontogenesis. *Russ. J. Dev. Biol.* **2018**, *49*, 61–78. [CrossRef]

7. Chaparro, C.; Gayraud, T.; De Souza, R.F.; Domingues, D.S.; Akaffou, S.S.; Vanzela, A.L.L.; De Kochko, A.; Rigoreau, M.; Crouzillat, D.; Hamon, S.; et al. Terminal-repeat retrotransposons with GAG domain in plant genomes: A new testimony on the complex world of transposable elements. *Genome Biol. Evol.* **2015**, *7*, 493–504. [CrossRef]

8. Wicker, T.; Sabot, F.; Hua-Van, A.; Bennetzen, J.L.; Capy, P.; Chalhoub, B.; Flavell, A.; Leroy, P.; Morgante, M.; Panaud, O.; et al. A unified classification system for eukaryotic transposable elements. *Nat. Rev. Genet.* **2007**, *8*, 973–982. [CrossRef]

9. Neumann, P.; Novák, P.; Hoštáková, N.; MacAs, J. Systematic survey of plant LTR-retrotransposons elucidates phylogenetic relationships of their polyprotein domains and provides a reference for element classification. *Mob. DNA* **2019**, *10*, 1–17. [CrossRef]

10. Loureiro, T.; Camacho, R.; Vieira, J.; Fonseca, N.A. Improving the performance of Transposable Elements detection tools. *J. Integr. Bioinform.* **2013**, *10*, 231. [CrossRef] [PubMed]

11. Nakano, F.K.; Mastelini, S.M.; Barbon, S.; Cerri, R. Improving Hierarchical Classification of Transposable Elements using Deep Neural Networks. In Proceedings of the Proceedings of the International Joint Conference on Neural Networks, Rio de Janeiro, Brazil, 8–13 July 2018; Volume 8–13 July.

12. Lewin, H.A.; Robinson, G.E.; Kress, W.J.; Baker, W.J.; Coddington, J.; Crandall, K.A.; Durbin, R.; Edwards, S.V.; Forest, F.; Gilbert, M.T.P.; et al. Earth BioGenome Project: Sequencing life for the future of life. *Proc. Natl. Acad. Sci. USA* **2018**, *115*, 4325–4333. [CrossRef]

13. Orozco-arias, S.; Isaza, G.; Guyot, R.; Tabares-soto, R. A systematic review of the application of machine learning in the detection and classi fi cation of transposable elements. *PeerJ* **2019**, *7*, 18311. [CrossRef] [PubMed]

14. Jurka, J.; Kapitonov, V.V.; Pavlicek, A.; Klonowski, P.; Kohany, O.; Walichiewicz, J. Repbase Update, a database of eukaryotic repetitive elements. *Cytogenet. Genome Res.* **2005**, *110*, 462–467. [CrossRef] [PubMed]

15. Cornut, G.; Choisne, N.; Alaux, M.; Alfama-Depauw, F.; Jamilloux, V.; Maumus, F.; Letellier, T.; Luyten, I.; Pommier, C.; Adam-Blondon, A.-F.; et al. RepetDB: A unified resource for transposable element references. *Mob. DNA* **2019**, *10*, 6.

16. Wicker, T.; Matthews, D.E.; Keller, B. TREP: A database for Triticeae repetitive elements 2002. Available online: http://botserv2.uzh.ch/kelldata/trep-db/pdfs/2002_TIPS.pdf (accessed on 24 May 2020).

17. Spannagl, M.; Nussbaumer, T.; Bader, K.C.; Martis, M.M.; Seidel, M.; Kugler, K.G.; Gundlach, H.; Mayer, K.F.X. PGSB PlantsDB: Updates to the database framework for comparative plant genome research. *Nucleic Acids Res.* **2015**, *44*, D1141–D1147. [CrossRef] [PubMed]

18. Du, J.; Grant, D.; Tian, Z.; Nelson, R.T.; Zhu, L.; Shoemaker, R.C.; Ma, J. SoyTEdb: A comprehensive database of transposable elements in the soybean genome. *BMC Genom.* **2010**, *11*, 113. [CrossRef]

19. Llorens, C.; Futami, R.; Covelli, L.; Domínguez-Escribá, L.; Viu, J.M.; Tamarit, D.; Aguilar-Rodríguez, J.; Vicente-Ripolles, M.; Fuster, G.; Bernet, G.P.; et al. The Gypsy Database (GyDB) of Mobile Genetic Elements: Release 2.0. *Nucleic Acids Res.* **2011**, *39*, 70–74. [CrossRef]

20. Pedro, D.L.F.; Lorenzetti, A.P.R.; Domingues, D.S.; Paschoal, A.R. PlaNC-TE: A comprehensive knowledgebase of non-coding RNAs and transposable elements in plants. *Database* **2018**, *2018*, bay078. [CrossRef]

21. Lorenzetti, A.P.R.; De Antonio, G.Y.A.; Paschoal, A.R.; Domingues, D.S. PlanTE-MIR DB: A database for transposable element-related microRNAs in plant genomes. *Funct. Integr. Genom.* **2016**, *16*, 235–242. [CrossRef]

22. Kamath, U.; De Jong, K.; Shehu, A. Effective automated feature construction and selection for classification of biological sequences. *PLoS ONE* **2014**, *9*, e99982. [CrossRef]

23. Nakano, F.K.; Martiello Mastelini, S.; Barbon, S.; Cerri, R. Stacking methods for hierarchical classification. In Proceedings of the 16th IEEE International Conference on Machine Learning and Applications, Cancun, Mexico, 18–21 December 2017; Volume 2018–Janua, pp. 289–296.

24. Nakano, F.K.; Pinto, W.J.; Pappa, G.L.; Cerri, R. Top-down strategies for hierarchical classification of transposable elements with neural networks. In Proceedings of the Proceedings of the International Joint Conference on Neural Networks, Anchorage, AK, USA, 14–19 May 2017; Volume 2017-May, pp. 2539–2546.

25. Ventola, G.M.M.; Noviello, T.M.R.; D'Aniello, S.; Spagnuolo, A.; Ceccarelli, M.; Cerulo, L. Identification of long non-coding transcripts with feature selection: A comparative study. *BMC Bioinform.* **2017**, *18*, 187. [CrossRef] [PubMed]

26. Rawal, K.; Ramaswamy, R. Genome-wide analysis of mobile genetic element insertion sites. *Nucleic Acids Res.* **2011**, *39*, 6864–6878. [CrossRef] [PubMed]

27. Zamith Santos, B.; Trindade Pereira, G.; Kenji Nakano, F.; Cerri, R. Strategies for selection of positive and negative instances in the hierarchical classification of transposable elements. In Proceedings of the Proceedings - 2018 Brazilian Conference on Intelligent Systems, Sao Paulo, Brazil, 22–25 October 2018; pp. 420–425.

28. Larrañaga, P.; Calvo, B.; Santana, R.; Bielza, C.; Galdiano, J.; Inza, I.; Lozano, J.A.; Armañanzas, R.; Santafé, G.; Pérez, A.; et al. Machine learning in bioinformatics. *Brief. Bioinform.* **2006**, *7*, 86–112. [CrossRef]

29. Mjolsness, E.; DeCoste, D. Machine learning for science: State of the art and future prospects. *Science (80-.)* **2001**, *293*, 2051–2055. [CrossRef] [PubMed]

30. Libbrecht, M.W.; Noble, W.S. Machine learning applications in genetics and genomics. *Nat. Rev. Genet.* **2015**, *16*, 321–332. [CrossRef] [PubMed]

31. Ceballos, D.; López-álvarez, D.; Isaza, G.; Tabares-Soto, R.; Orozco-Arias, S.; Ferrin, C.D. A Machine Learning-based Pipeline for the Classification of CTX-M in Metagenomics Samples. *Processes* **2019**, *7*, 235. [CrossRef]

32. Loureiro, T.; Camacho, R.; Vieira, J.; Fonseca, N.A. Boosting the Detection of Transposable Elements Using Machine Learning. In *7th International Conference on Practical Applications of Computational Biology & Bioinformatics*; Springer: Heidelberg, Germany, 2013; pp. 85–91.

33. Santos, B.Z.; Cerri, R.; Lu, R.W. A New Machine Learning Dataset for Hierarchical Classification of Transposable Elements. In Proceedings of the XIII Encontro Nacional de Inteligência Artificial-ENIAC, Sao Paulo, Brazil, 9–12 October 2016.

34. Schietgat, L.; Vens, C.; Cerri, R.; Fischer, C.N.; Costa, E.; Ramon, J.; Carareto, C.M.A.; Blockeel, H. A machine learning based framework to identify and classify long terminal repeat retrotransposons. *PLoS Comput. Biol.* **2018**, *14*, e1006097. [CrossRef] [PubMed]

35. Ma, C.; Zhang, H.H.; Wang, X. Machine learning for Big Data analytics in plants. *Trends Plant Sci.* **2014**, *19*, 798–808. [CrossRef] [PubMed]

36. Müller, A.C.; Guido, S. *Introduction to Machine Learning with Python: A Guide for Data Scientists*; O'Reilly Media, Inc.: Sebastopol, CA, USA, 2016.

37. Liu, Y.; Zhou, Y.; Wen, S.; Tang, C. A Strategy on Selecting Performance Metrics for Classifier Evaluation. *Int. J. Mob. Comput. Multimed. Commun.* **2014**, *6*, 20–35. [CrossRef]

38. Eraslan, G.; Avsec, Ž.; Gagneur, J.; Theis, F.J. Deep learning: New computational modelling techniques for genomics. *Nat. Rev. Genet.* **2019**, *20*, 389–403. [CrossRef]

39. Tsafnat, G.; Setzermann, P.; Partridge, S.R.; Grimm, D. Computational inference of difficult word boundaries in DNA languages. In *Proceedings of the ACM International Conference Proceeding Series; Barcelona*; Kyranova Ltd, Center for TeleInFrastruktur: Barcelona, Spain, 2011.

40. Sokolova, M.; Lapalme, G. A systematic analysis of performance measures for classification tasks. *Inf. Process. Manag.* **2009**, *45*, 427–437. [CrossRef]

41. Girgis, H.Z. Red: An intelligent, rapid, accurate tool for detecting repeats de-novo on the genomic scale. *BMC Bioinform.* **2015**, *16*, 1–19. [CrossRef] [PubMed]

42. Su, W.; Gu, X.; Peterson, T. TIR-Learner, a New Ensemble Method for TIR Transposable Element Annotation, Provides Evidence for Abundant New Transposable Elements in the Maize Genome. *Mol. Plant* **2019**, *12*, 447–460. [CrossRef] [PubMed]

43. Arango-López, J.; Orozco-Arias, S.; Salazar, J.A.; Guyot, R. Application of Data Mining Algorithms to Classify Biological Data: The Coffea canephora Genome Case. In *Colombian Conference on Computing*; Springer: Cartagena, Colombia, 2017; Volume 735, pp. 156–170. ISBN 9781457720819.

44. Hesam, T.D.; Ali, M.-N. Mining biological repetitive sequences using support vector machines and fuzzy SVM. *Iran. J. Chem. Chem. Eng.* **2010**, *29*, 1–17.

45. Abrusán, G.; Grundmann, N.; Demester, L.; Makalowski, W. TEclass - A tool for automated classification of unknown eukaryotic transposable elements. *Bioinformatics* **2009**, *25*, 1329–1330. [CrossRef]

46. Castellanos-Garzón, J.A.; Díaz, F. Boosting the Detection of Transposable Elements UsingMachine Learning. *Adv. Intell. Syst. Comput.* **2013**, *222*, 15–22.

47. Da Cruz, M.H.P.; Saito, P.T.M.; Paschoal, A.R.; Bugatti, P.H. Classification of Transposable Elements by Convolutional Neural Networks. In *Lecture Notes in Computer Science*; Springer International Publishing: New York, NY, USA, 2019; Volume 11509, pp. 157–168. ISBN 9783030209155.

48. Kitchenham, B.; Charters, S. *Guidelines for Performing Systematic Literature Reviews in Software Engineering*; Version 2.3 EBSE Technical Report EBSE-2007-01; Department of Computer Science University of Durham: Durham, UK, 2007.

49. Marchand, M.; Shawe-Taylor, J. The set covering machine. *J. Mach. Learn. Res.* **2002**, *3*, 723–746.

50. Caruana, R.; Niculescu-Mizil, A. An empirical comparison of supervised learning algorithms. In Proceedings of the Proceedings of the 23rd International Conference on Machine Learning, New York, NY, USA, 25–29 June 2006; pp. 161–168.

51. Schnable, P.S.; Ware, D.; Fulton, R.S.; Stein, J.C.; Wei, F.; Pasternak, S.; Liang, C.; Zhang, J.; Fulton, L.; Graves, T.A.; et al. The B73 Maize Genome: Complexity, Diversity, and Dynamics. *Science (80-.)* **2009**, *326*, 1112–1115. [CrossRef]

52. Choulet, F.; Alberti, A.; Theil, S.; Glover, N.; Barbe, V.; Daron, J.; Pingault, L.; Sourdille, P.; Couloux, A.; Paux, E.; et al. Structural and functional partitioning of bread wheat chromosome 3B. *Science (80-.)* **2014**, *345*, 1249721. [CrossRef]

53. Paterson, A.H.; Bowers, J.E.; Bruggmann, R.; Dubchak, I.; Grimwood, J.; Gundlach, H.; Haberer, G.; Hellsten, U.; Mitros, T.; Poliakov, A.; et al. The Sorghum bicolor genome and the diversification of grasses. *Nature* **2009**, *457*, 551–556. [CrossRef]

54. Denoeud, F.; Carretero-Paulet, L.; Dereeper, A.; Droc, G.; Guyot, R.; Pietrella, M.; Zheng, C.; Alberti, A.; Anthony, F.; Aprea, G.; et al. The coffee genome provides insight into the convergent evolution of caffeine biosynthesis. *Science (80-.)* **2014**, *345*, 1181–1184. [CrossRef] [PubMed]

55. Orozco-arias, S.; Liu, J.; Id, R.T.; Ceballos, D.; Silva, D.; Id, D.; Ming, R.; Guyot, R. Inpactor, Integrated and Parallel Analyzer and Classifier of LTR Retrotransposons and Its Application for Pineapple LTR Retrotransposons Diversity and Dynamics. *Biology (Basel)* **2018**, *7*, 32. [CrossRef] [PubMed]

56. Jaiswal, A.K.; Krishnamachari, A. Physicochemical property based computational scheme for classifying DNA sequence elements of Saccharomyces cerevisiae. *Comput. Biol. Chem.* **2019**, *79*, 193–201. [CrossRef] [PubMed]

57. Yu, N.; Guo, X.; Gu, F.; Pan, Y. DNA AS X: An information-coding-based model to improve the sensitivity in comparative gene analysis. In Proceedings of the International Symposium on Bioinformatics Research and Applications, Norfolk, VA, USA, 6–9 June 2015; pp. 366–377.

58. Nair, A.S.; Sreenadhan, S.P. A coding measure scheme employing electron-ion interaction pseudopotential (EIIP). *Bioinformation* **2006**, *1*, 197. [PubMed]

59. Akhtar, M.; Epps, J.; Ambikairajah, E. Signal processing in sequence analysis: Advances in eukaryotic gene prediction. *IEEE J. Sel. Top. Signal Process.* **2008**, *2*, 310–321. [CrossRef]

60. Kauer, G.; Blöcker, H. Applying signal theory to the analysis of biomolecules. *Bioinformatics* **2003**, *19*, 2016–2021. [CrossRef]

61. Rosen, G.L. Signal Processing for Biologically-Inspired Gradient Source Localization and DNA Sequence Analysis. Ph.D. Thesis, Georgia Institute of Technology, Atlanta, GA, USA, 12 July 2006.

62. Tabares-soto, R.; Orozco-Arias, S.; Romero-Cano, V.; Segovia Bucheli, V.; Rodríguez-Sotelo, J.L.; Jiménez-Varón, C.F. A comparative study of machine learning and deep learning algorithms to classify cancer types based on microarray gene expression. *Peerj Comput. Sci.* **2020**, *6*, 1–22. [CrossRef]

63. Pedregosa, F.; Varoquaux, G.; Gramfort, A.; Michel, V.; Thirion, B.; Grisel, O.; Blondel, M.; Prettenhofer, P.; Weiss, R.; Dubourg, V.; et al. Scikit-learn: Machine Learning in Python. *J. Mach. Learn. Res.* **2011**, *12*, 2825–2830.

64. Chen, L.; Zhang, Y.-H.; Huang, G.; Pan, X.; Wang, S.; Huang, T.; Cai, Y.-D. Discriminating cirRNAs from other lncRNAs using a hierarchical extreme learning machine (H-ELM) algorithm with feature selection. *Mol. Genet. Genom.* **2018**, *293*, 137–149. [CrossRef] [PubMed]

65. Yu, N.; Yu, Z.; Pan, Y. A deep learning method for lincRNA detection using auto-encoder algorithm. *BMC Bioinform.* **2017**, *18*, 511. [CrossRef] [PubMed]

66. Smith, M.A.; Seemann, S.E.; Quek, X.C.; Mattick, J.S. DotAligner: Identification and clustering of RNA structure motifs. *Genome Biol.* **2017**, *18*, 244. [CrossRef] [PubMed]

67. Segal, E.S.; Gritsenko, V.; Levitan, A.; Yadav, B.; Dror, N.; Steenwyk, J.L.; Silberberg, Y.; Mielich, K.; Rokas, A.; Gow, N.A.R.; et al. Gene Essentiality Analyzed by In Vivo Transposon Mutagenesis and Machine Learning in a Stable Haploid Isolate of Candida albicans. *MBio* **2018**, *9*, e02048-18. [CrossRef] [PubMed]

68. Brayet, J.; Zehraoui, F.; Jeanson-Leh, L.; Israeli, D.; Tahi, F. Towards a piRNA prediction using multiple kernel fusion and support vector machine. *Bioinformatics* **2014**, *30*, i364–i370. [CrossRef] [PubMed]

69. Ashlock, W.; Datta, S. Distinguishing endogenous retroviral LTRs from SINE elements using features extracted from evolved side effect machines. *IEEE/ACM Trans. Comput. Biol. Bioinforma.* **2012**, *9*, 1676–1689. [CrossRef]

70. Zhang, Y.; Babaian, A.; Gagnier, L.; Mager, D.L. Visualized Computational Predictions of Transcriptional Effects by Intronic Endogenous Retroviruses. *PLoS ONE* **2013**, *8*, e71971. [CrossRef] [PubMed]

71. Douville, C.; Springer, S.; Kinde, I.; Cohen, J.D.; Hruban, R.H.; Lennon, A.M.; Papadopoulos, N.; Kinzler, K.W.; Vogelstein, B.; Karchin, R. Detection of aneuploidy in patients with cancer through amplification of long interspersed nucleotide elements (LINEs). *Proc. Natl. Acad. Sci. USA* **2018**, *115*, 1871–1876. [CrossRef]

72. Rishishwar, L.; Mariño-Ramírez, L.; Jordan, I.K. Benchmarking computational tools for polymorphic transposable element detection. *Brief. Bioinform.* **2017**, *18*, 908–918. [CrossRef]

73. Youden, W.J. Index for rating diagnostic tests. *Cancer* **1950**, *3*, 32–35. [CrossRef]

74. Gao, D.; Abernathy, B.; Rohksar, D.; Schmutz, J.; Jackson, S.A. Annotation and sequence diversity of transposable elements in common bean (Phaseolus vulgaris). *Front. Plant Sci.* **2014**, *5*, 339. [CrossRef]

75. Jiang, N. Overview of Repeat Annotation and De Novo Repeat Identification. In *Plant Transposable Elements*; Humana Press: Totowa, NJ, USA, 2013; pp. 275–287.

76. Garbus, I.; Romero, J.R.; Valarik, M.; Vanžurová, H.; Karafiátová, M.; Cáccamo, M.; Doležel, J.; Tranquilli, G.; Helguera, M.; Echenique, V. Characterization of repetitive DNA landscape in wheat homeologous group 4 chromosomes. *BMC Genom.* **2015**, *16*, 375. [CrossRef]

77. Eickbush, T.H.; Jamburuthugoda, V.K. The diversity of retrotransposons and the properties of their reverse transcriptases. *VIRUS Res.* **2008**, *134*, 221–234. [CrossRef] [PubMed]

78. Negi, P.; Rai, A.N.; Suprasanna, P. Moving through the Stressed Genome: Emerging Regulatory Roles for Transposons in Plant Stress Response. *Front. Plant Sci.* **2016**, *7*, 1448. [CrossRef] [PubMed]

79. Bousios, A.; Minga, E.; Kalitsou, N.; Pantermali, M.; Tsaballa, A.; Darzentas, N. MASiVEdb: The Sirevirus Plant Retrotransposon Database. *BMC Genom.* **2012**, *13*, 158. [CrossRef] [PubMed]

80. Naresh, E.; Kumar, B.P.V.; Shankar, S.P. Others Impact of Machine Learning in Bioinformatics Research. In *Statistical Modelling and Machine Learning Principles for Bioinformatics Techniques, Tools, and Applications*; Springer: Singapore, 2020; pp. 41–62.

81. Yue, T.; Wang, H. Deep Learning for Genomics: A Concise Overview. *arXiv* **2018**, arXiv:1802.008101–40.

82. Soueidan, H.; Nikolski, M. Machine learning for metagenomics: Methods and tools. *arXiv* **2015**, arXiv:1510.06621. 2015. [CrossRef]

83. Captur, G.; Heywood, W.E.; Coats, C.; Rosmini, S.; Patel, V.; Lopes, L.R.; Collis, R.; Patel, N.; Syrris, P.; Bassett, P.; et al. Identification of a multiplex biomarker panel for Hypertrophic Cardiomyopathy using quantitative proteomics and machine learning. *Mol. Cell. Proteom.* **2020**, *19*, 114–127. [CrossRef]

84. Loureiro, T.; Fonseca, N.; Camacho, R. Application of Machine Learning Techniques on the Discovery and Annotation of Transposons in Genomes. Master's Thesis, Faculdade De Engenharia, Universidade Do Porto, Porto, Portugal, 2012.

85. Guyot, R.; Darré, T.; Dupeyron, M.; de Kochko, A.; Hamon, S.; Couturon, E.; Crouzillat, D.; Rigoreau, M.; Rakotomalala, J.-J.; Raharimalala, N.E.; et al. Partial sequencing reveals the transposable element composition of Coffea genomes and provides evidence for distinct evolutionary stories. *Mol. Genet. Genom.* **2016**, *291*, 1979–1990. [CrossRef]

86. Piegu, B.; Guyot, R.; Picault, N.; Roulin, A.; Saniyal, A.; Kim, H.; Collura, K.; Brar, D.S.; Jackson, S.; Wing, R.A.; et al. Doubling genome size without polyploidization: Dynamics of retrotransposition-driven genomic expansions in Oryza australiensis, a wild relative of rice. *Genome Res.* **2006**, *16*, 1262–1269. [CrossRef]

87. Ming, R.; VanBuren, R.; Wai, C.M.; Tang, H.; Schatz, M.C.; Bowers, J.E.; Lyons, E.; Wang, M.-L.; Chen, J.; Biggers, E.; et al. The pineapple genome and the evolution of CAM photosynthesis. *Nat. Genet.* **2015**, *47*, 1435–1442. [CrossRef]

 *processes*

Article

An Adjective Selection Personality Assessment Method Using Gradient Boosting Machine Learning

Bruno Fernandes [1],*, Alfonso González-Briones [2,3], Paulo Novais [1], Miguel Calafate [1], Cesar Analide [1] and José Neves [1]

[1] Department of Informatics, ALGORITMI Centre, University of Minho, 4704-553 Braga, Portugal; pjon@di.uminho.pt (P.N.); calafateds@gmail.com (M.C.); analide@di.uminho.pt (C.A.); jneves@di.uminho.pt (J.N.)

[2] Research Group on Agent-Based, Social and Interdisciplinary Applications (GRASIA), Complutense University of Madrid, 28040 Madrid, Spain; alfonsogb@ucm.es

[3] BISITE Research Group, University of Salamanca, Edificio I+D+i, 37007 Salamanca, Spain

* Correspondence: bruno.fmf.8@gmail.com

Received: 16 April 2020; Accepted: 18 May 2020; Published: 21 May 2020

Abstract: Goldberg's 100 Unipolar Markers remains one of the most popular ways to measure personality traits, in particular, the Big Five. An important reduction was later preformed by Saucier, using a sub-set of 40 markers. Both assessments are performed by presenting a set of markers, or adjectives, to the subject, requesting him to quantify each marker using a 9-point rating scale. Consequently, the goal of this study is to conduct experiments and propose a shorter alternative where the subject is only required to identify which adjectives describe him the most. Hence, a web platform was developed for data collection, requesting subjects to rate each adjective and select those describing him the most. Based on a Gradient Boosting approach, two distinct Machine Learning architectures were conceived, tuned and evaluated. The first makes use of regressors to provide an exact score of the Big Five while the second uses classifiers to provide a binned output. As input, both receive the one-hot encoded selection of adjectives. Both architectures performed well. The first is able to quantify the Big Five with an approximate error of 5 units of measure, while the second shows a micro-averaged f1-score of 83%. Since all adjectives are used to compute all traits, models are able to harness inter-trait relationships, being possible to further reduce the set of adjectives by removing those that have smaller importance.

Keywords: Machine Learning; personality assessment; gradient boosting; Affective Computing

---

1. Introduction

People react differently when experiencing the same situations. This behavioural diversity may be due to one's experience, knowledge or even personality. Indeed, several studies have already established a relationship between a person's personality and aggressive reactions [1], work performance [2] or infidelity [3], just to name a few. Semantically, personality may be defined as a set of characteristics that refer to individual differences in ways of thinking, feeling and behaving [4]. Personality has a great impact in the the way we live our lives, either by the way we behave, feel or interact with others. Hence, there has always been great interest in model, or quantify, a person's personality using either qualitative or quantitative metrics. Nowadays, there are several accepted tests that allow a psychological assessment of a person. Such tests may be performed in the scope of psychology appointments, job interviews or psychometric evaluations. These tests are mainly conducted by trained professionals that are able to properly interpret their results.

1.1. Personality Assessment

Personality assessment is a well defined process that can help unveil how a person may react to different, unexpected, situations [5]. Several personality tests are available such as HEXACO-60 [6], Myers-Briggs Type Indicator [7], the Enneagram of Personality [8] and NEO-personality-inventory (NEO-PI-R) [9]. Goldberg's 100 Unipolar Markers' Test [10] is yet another test that consists of a total of 100 adjectives, or markers, that the subject must rate on how they relate to each adjective, with 1 being *Extremely Inaccurate* and 9 *Extremely Accurate*. Among the full set of markers one may find adjectives such as *talkative, sympathetic, careless, envious* or *deep*. Goldberg's test allows one to measure five domains, in particular, *Surgency, Agreeableness, Conscientiousness, Emotional Stability* and *Intellect*. Different domains have also been proposed. The OCEAN model, on the other hand, consists of the following five factors [11,12]:

- *Openness*: related to one's curiosity, imagination and openness to new experiences. Higher values usually emerge on people that enjoy new adventures and ideas. On the other hand, lower values tend to emerge on more conservative people;
- *Conscientiousness*: related to self-discipline, being careful and diligent, organised and consistent, pursuing long-term goals. Less conscientious people tend to be more spontaneous and imaginative;
- *Extraversion*: related to a state where a person seeks stimulation from being with others instead of being alone. Extroverted people tend to be energetic and talkative, while introverted ones are reserved and prefer not to be the centre of attention;
- *Agreeableness*: related to behavioural characteristics such as being kind and sympathetic. Agreeable people tend to be friendly, cooperative and empathetic. Non-agreeable people are less cooperative, and more competitive and suspicious;
- *Neuroticism* (opposite of *Stability*): related to being moody and showing signs of emotional instability. Neurotic people tend to be stressed and nervous. Non-neurotic people tend to be calmer and more emotionally stable [12].

Goldberg's test consist of 100 unipolar markers that must be quantified by a subject. An important reduction to the set of markers was performed by Gerard Saucier with The Mini-Marker test, using a sub-set of 40 markers to assess the Big Five with an acceptable performance, leading to the use of less difficult markers and lower inter-scale correlations [13]. Saucier's test uses the same rating scale, being made of five disjoint-sets of eight unipolar markers each:

- *Intellect or Openness* trait is made of six positively weighted adjectives (*intellectual, creative, complex, imaginative, philosophical* and *deep*) and two negative ones (*average* and *ordinary*);
- *Conscientiousness* trait is made of four positively weighted adjectives (*systematic, practical, efficient* and *orderly*) and four negative ones (*disorganised, careless, inefficient* and *sloppy*);
- *Extraversion* trait is made of four positively weighted adjectives (*extraverted, talkative, energetic* and *bold*) and four negative ones (*shy, quiet, withdrawn* and *bashful*);
- *Agreeableness* trait is made of four positively weighted adjectives (*kind, cooperative, sympathetic* and *warm*) and four negative ones (*cold, harsh, rude* and *distant*);
- *Emotional Stability* trait is made of two positively weighted adjectives (*relaxed* and *mellow*) and six negative ones (*moody, temperamental, envious, fretful, jealous* and *touchy*).

1.2. Machine Learning for Personality Assessment

During these last years, Machine Learning (ML) has been raising to prominence. In fact, the use of ML models to predict personality traits has gain significant popularity within the field of Affective Computing, with several studies having already engaged on conceiving ML models for personality assessment [12,14–16]. In 2017, Majumder et al. conceived and evaluated Deep Learning (DL) models to assess personality from text. They conceived and fit a total of five artificial neural networks (ANN), one for each of the Big Five personality traits. All networks had the same architecture, with each ANN behaving as a binary classifier to predict whether the trait was positive or negative [14]. As dataset the authors used James Pennebaker and Laura King's stream-of-consciousness essay dataset, which contains 2468 anonymous essays tagged with the binary value for each of the Big Five [17]. This dataset seems, however, to be currently unavailable. In fact, several datasets containing anonymized psychological assessments seem to have been locked, or closed, such as the one provided by the myPersonality platform (myPersonality.org), a platform that made available a dataset containing textual social media data and from where several studies emerged, being essentially focused on modelling personality traits based on language-based information [18,19].

In a slightly different domain, in 2017, Yu and Markov conceived and evaluated several DL models to learn suitable data representation for personality assessment, using facebook status update data. This dataset consisted of raw text, user's information and standard Big Five labels, which were obtained using self-assessment questionnaires [15]. In fact, it is possible to find several studies focused on inferring personality based on social media feeds. For instance, Kosinski et al. (2014) focused on examining how an individual's personality manifests in his/her online behaviour, in particular, the website he/she visits and his/her Facebook activity. The expectation is that web activity combined with social media data may bring unbiased insights, since social media feeds may carry an intention of self-enhancement and positivity [12]. The used dataset was obtained from myPersonality. The obtained results showed psychologically meaningful links between individuals' personalities, website preferences and social media data. The potential applications of these works are essentially related with targeted advertising and personalised recommender systems, which take into consideration one's personality to deliver useful content.

In 2012, Sumner et al., based on Twitter use, focused on identifying signals of the Dark Triad, i.e., the anti-social traits of *Narcissism*, *Machiavellianism* and *Psychopathy*. Almost three thousand Twitter users, from 89 countries, participated in the study, with an in-built Twitter application being developed to collect self-reported ratings on the Short Dark Triad questionnaire, which measures the anti-social traits, and the Ten Item Personality Inventory (TIPI) test, which measures the Big Five. The authors conclude that even though possible to examine large groups of people, the conceived ML models behave poorly when applied to individuals, being imprecise when predicting Dark Triad traits just from Twitter activity [16].

Another study, performed by Cerasa et al. (2018), focused on conceiving and evaluating ML models to identify individuals with gambling disorder. To build the dataset, a set of healthy and sick individuals were asked to perform the NEO-PI-R test, an operationalization of the five factor model. The authors employed Classification and Regression Trees (CART) achieving interesting performances evaluated using the area under the curve (AUC). In fact, the best candidate model was able to identify individuals with gambling disorder with an AUC of approximately 77% [20].

On the other hand, studies have been performed where audio and video data are used by DL-based models to predict personality [21]. One study, performed by Levitan et al. (2016), focused on the automatic identification of traits such as gender, deception and personality using acoustic-prosodic and lexical features [22]. In particular, the authors focused on automatic detection of deception. The authors used Columbia deception corpus, which consists of deceptive and non-deceptive speech from standard American-English and Mandarin-Chinese native speakers, including more than one hundred hours of speech with self-identified truth/lie labels [23]. The authors then collected demographic data from each subject and administered a NEO-FFI personality test to access the Big Five. Each trait was binned as a three-class classification problem (*low*, *medium* and *high*), which created an highly unbalanced dataset since the majority of subjects fell into the *medium* class. Hence, to compare models' performances the authors used f-scores to obtain a meaningful comparison. Several ML models and feature sets were experimented, with AdaBoost and Random Forests being the best performing classifiers for personality assessment [22].

Another study, performed by Gurpinar et al. (2016), focused in using DL to predict the Big Five of faces appearing in videos [24]. The authors employed transfer-learning and Convolutional

Neural Networks to extract facial expressions, as well as ambient information. The conceived models were evaluated on the *ChaLearn Challenge Dataset on First Impression Recognition*, which consists of ten thousand clips collected from more that five thousand YouTube videos. The label of each clip corresponds to the Big Five personality traits of the person appearing in that clip. Their best candidate model achieved an accuracy of over 90% on the test set [24].

It is also usual to find the use of different data sources combined through means of data fusion for personality assessment. Indeed, personality assessment from multi-modal data has been assuming a greater importance in the computer vision field [25]. For instance, Gucluturk et al. (2017), aimed to analyse what features are used by personality trait assessment models when making predictions, conducting several experiments that characterised audio and visual information that drive such predictions [25]. On the other hand, Zhang et al. (2016) proposed a Deep Bimodal Regression framework to capture rich information from both the visual and audio aspects of videos, winning the *ChaLearn Looking at People* challenge. Convolutional Neural Networks were conceived to exploit visual cues, while linear regressors where used for audio [26].

1.3. Hypothesis and Paper Structure

Many studies have already engaged on using ML or DL for personality assessment using images, videos, audio or text. However, to the best of our knowledge, we are the first to apply ML to reduce the complexity of a test. In fact, the working hypothesis is that it is possible to use ML-based modes to further reduce Saucier's Mini-Marker to a "game of words" where the subject, instead of rating forty adjectives, only has to select those he relates the most, removing the need to rate adjectives. The proposed Adjective Selection to Assess Personality (ASAP) method replaces the entire process of rating adjectives by an adjective selection process. The goal is to reduce the complexity of tests, the time it takes to perform a test, and to make the test more attractive and easier to implement in current and future technological platforms. Hence, this study aims to conceive, tune and evaluate two distinct Gradient Boosting ML architectures to quantify an individual's personality based on his/her choice of adjectives. Due to the non-availability of data, a web platform was developed and place online, being responsible for the entire data collection process. To conduct experiments on non-data scarce environments, data augmentation techniques were designed and implemented to produce a second dataset, which was also evaluated.

The remainder of this paper is structured as follows, viz. Section 2 describes the material and methods, in particular the developed platform for data collection, data exploration, the implemented data augmentation techniques as well as the conceived ML architectures, the experimental setup and the conducted experiments. Section 3 summarises the obtained results, providing a concise description of the experimental results and their interpretation. Section 4 presents and discusses the results and their interpretation in the perspective of previous studies and of the working hypothesis, depicting the main conclusions and pointing future research directions.

2. Materials and Methods

Due to the non-availability of data and the particularities of the proposed ASAP method, we were required to develop a web platform for data collection, requesting subjects to rate adjectives and select those describing them the most. This allowed us to build a dataset containing self-reported ratings on Saucier's Mini-Marker test, the corresponding values of the Big Five as well as the adjectives selected by the subjects. The next lines describe in detail the developed platform, exploring and explaining the collected dataset and the implemented data augmentation techniques. It also details the conceived ML architectures and the experimental setup.

2.1. Dataset

The dataset used in this study is available, in its raw state, in an online repository (https://github.com/brunofmf/Datasets4SocialGood), under a MIT license.

2.1.1. Data Collection

To bring this study to a fruitful conclusion, we were required to collect a dataset from where we could derive conclusions. Hence, a platform was conceived and made available online (http://crowdsensing.di.uminho.pt/). The platform displays all 40 adjectives used by Saucier's Mini-Marker test, asking the subject to rate each one. It also allows the subject to select a set of adjectives that describe him the most. Figure 1 depicts the main page of the conceived platform. The subject can then get the test results and obtain the value of each personality trait.

The platform provides a rationale to explain the subject how he/she is contributing to the study. No personal data are stored neither it is possible to link subjects to their answers - only information about age, genre and language are stored, and only if the user explicitly provides it. The platform is available online and any person can access and use it. It was published online on 21 September 2018. The platform was shared among a diversified population, using social media and university's mailing lists. Data was also collected in person, which allowed us to increment the dataset size with records containing both the ratings and the selected list of adjectives.

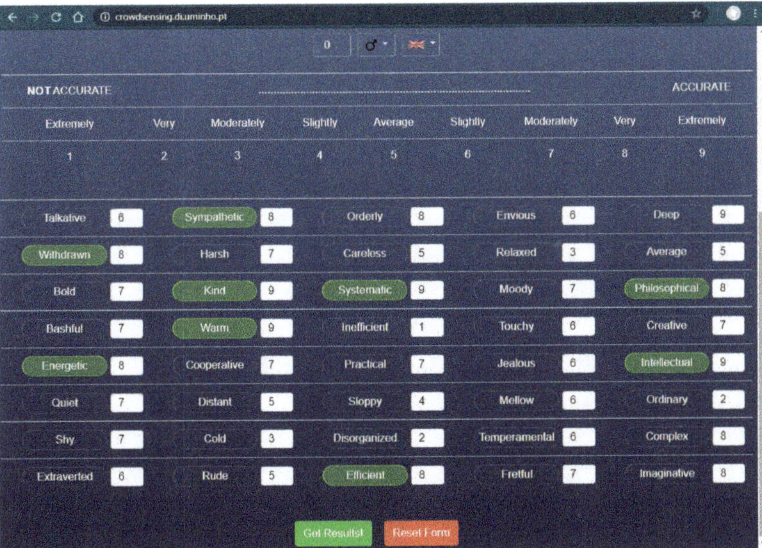

Figure 1. Platform for data collection allowing the subject to perform Saucier's Mini-Marker test and, at the same time, select a set of adjectives that describe him the most.

To facilitate the data collection process, the developed platform allows subjects to perform Saucier's Mini-Markers in three distinct languages. All translations were performed by three Portuguese and Spanish native speakers fluent in English, all university professors. It should also be highlighted that this study does not aim to examine the psychometric properties of the Portuguese or Spanish versions neither to provide sound validity evidence for the performed translations (even though *Tau-Equivalent* estimates of score reliability are later examined). The assumption is that ML models are able to quantify or qualify the traits without requiring any contextual information about region, genre, language or age of the subjects.

2.1.2. Data Exploration

The collected dataset contains 255 observations. Each observation is made of 50 features, viz, *age, genre, language,* 40 *adjectives,* 5 *personality traits,* the *selected adjectives* and the *creation date.* The features *age, adjectives* and *personality traits* are integers. The *genre* is a binary attribute and *language* is either *es*, *en* or *pt*. On the other hand, the *selected adjectives* feature consists of a string where the selected adjectives are comma separated. Table 1 presents all available features in the collected dataset.

Table 1. Features available in the collected dataset.

#	Feature	#	Feature	#	Feature
1	age	18	philosophical	35	cold
2	genre	19	bashful	36	disorganized
3	language	20	warm	37	temperamental
4	talkative	21	inefficient	38	complex
5	sympathetic	22	touchy	39	extraverted
6	orderly	23	creative	40	rude
7	envious	24	energetic	41	efficient
8	deep	25	cooperative	42	fretful
9	withdrawn	26	practical	43	imaginative
10	harsh	27	jealous	44	extraversion_trait
11	careless	28	intellectual	45	agreeableness_trait
12	relaxed	29	quiet	46	conscientiousness_trait
13	average	30	distant	47	stability_trait
14	bold	31	sloppy	48	openess_trait
15	kind	32	mellow	49	selected_attr
16	systematic	33	ordinary	50	creation_date
17	moody	34	shy		

In the final dataset, 159 observations have the *selected_attr* feature filled with the selected adjectives. On the other hand, 96 observations only have the adjectives' ratings. A few observations have adjectives rated with the value 0. 200 observations belong to male subjects, while 55 belong to female ones. Only two languages were used: 220 observations were done in Portuguese while 35 were done in English. More than 90% of the observations were collected in 2019. The mean age value is of 30.1 years.

Adjectives with lower mean value are essentially related to negative ones such as *rude*, with 3.13, *inefficient*, with 3.26, and *ordinary*, with 3.28. The adjectives that have higher mean value are *kind*, with 6.004, *imaginative*, with 6, and *cooperative*, with 5.73. Mean standard deviation of the 40 adjectives is 2.5, with the lower value being 0 and the maximum 9. Mean skewness is of 0.03, representing a symmetrical distribution. Mean kurtosis is of -0.98, representing a somewhat "light-tailed" dataset in regard to the 40 adjectives. In regard to the Big Five (Table 2), the one having lower mean value is *Extraversion*, with *Agreeableness* being the one with higher mean value. Mean standard deviation of all traits is of approximately 10 units of measure. The coefficient alpha for the forty items is of 0.82 [27]. For each individual trait, the *Tau-Equivalent* estimates of score reliability are lower, specially for the *Stability* factor. Except for the *selected_attr* feature, no missing values are present in the dataset.

With all features assuming a non-Gaussian distribution (under the Kolmogorov-Smirnov test with $p < 0.05$), the non-parametric Spearman's rank correlation coefficient was used. A few pairs of correlated features, in the form (*trait, adjective*), appear in the dataset. This is in line with expectations since the Big Five are mathematically based on the adjectives. Higher correlations appear for the pairs (*Agreeableness, Warm*), (*Conscientiousness, Efficient*), (*Openess, Complex*) and (*Extraversion, Extraverted*).

The *selected_attr* feature consists of a string where adjectives are separated by commas. An example of a valid value would be "*Talkative, Sympathetic, Kind, Energetic, Jealous, Intellectual, Extraverted, Efficient, Fretful*". From all 159 observations that have the *selected_attr* feature filled, 157 are unique values meaning that only three subjects chose the same adjectives. Interestingly, all adjectives were selected at least once. In fact, the least selected adjectives were *ordinary*, which was selected 14 times, *touchy*, 18 times, *rude*, 19 times, *cold* and *fretful*, 23 times. These are, essentially, adjectives with negative connotation. On the opposite spectrum, *kind* was selected 67 times, *imaginative*, 59 times, *sympathetic*, 58 times, *creative*, 57 times, and *withdrawn*, 56 times (Figure 2). Excluding those who opt not to select adjectives, 10 subjects only chose one adjective to describe themselves, while 14 subjects selected fifteen, or more, adjectives. The mean value is of approximately ten selected adjectives per subject.

Table 2. Descriptive statistics for the Big Five.

	Openness	Conscientiousness	Extraversion	Agreeableness	Stability
N° of Items	8	8	8	8	8
Mean	44.976	46.476	39.428	47.148	46.140
Median	46	47	40	48	46
Standard Deviation	10.315	10.547	10.017	10.056	8.860
Skewness	−0.212	−0.207	−0.135	−0.216	−0.047
Kurtosis	−0.260	−0.492	−0.344	−0.328	−0.484
Coefficient alpha	0.62	0.61	0.56	0.58	0.42

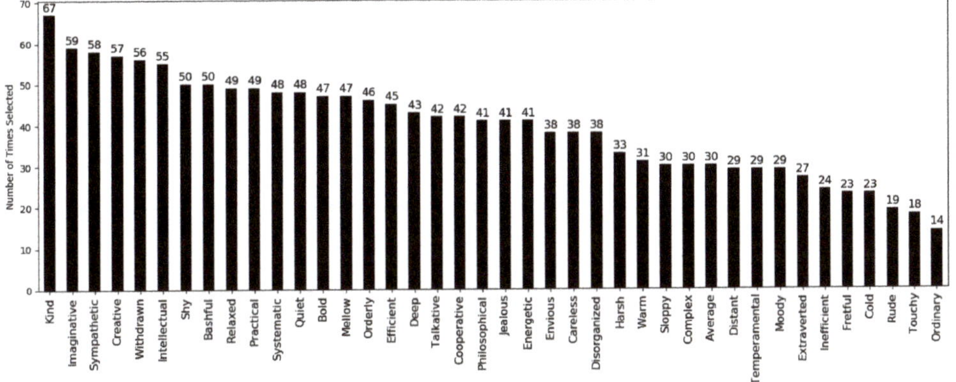

Figure 2. Number of times each adjective was selected.

Approximately 38% of the total number of observations do not have adjectives selected. To overcome this issue, it becomes important to understand the relation between selecting an adjective and its respective rating. For instance, considering all subjects that selected the adjective *efficient*, the mean rating of that same adjective is of 5.794. On the other hand, the mean rating of the *sloopy* adjective considering all subjects that selected that adjective is of 8. This tells us that *sloopy* tends to be selected when receiving higher ratings. On the other hand, *efficient* is selected even with average ratings. The overall mean, 7.448, tells us that, as expected, adjectives tend to be selected when receiving high values. Figure 3 depicts the mean rating values to set an adjective as selected.

To discover relations between the selected adjectives, a ML and a pattern mining method, entitled as Association Rules Learning (ARL), was applied. ARL does not consider the order of the items, neither extract individual's preference, but, instead, looks for frequent itemsets. The goal is to find associations and correlations between adjectives that were selected to describe subjects. In particular, the APRIORI algorithm was used to analyse the list of selected adjectives, and provide rules in the form *Antecedent -> Consequent*, where -> may be read as "implies". To find these rules, three distinct metrics were used: *Support*, which gives an idea of how frequent an itemset is in all existing transactions, helping identifying rules worth considering; *Confidence*, an indication of how often a rule has been found to be true; and *Lift*, which measures how much better the rule is at predicting the presence of an adjective compared to just relying on the raw probability of the adjective in the dataset. The returned rules go both ways, i.e., if *A* implies *B* then the reverse is also true. Table 3 presents all rules with a support value higher than 0.15. In fact, the support value was tuned in order to find a representative set of rules. Such a lower support value tells us that rules tend to be less frequent than expected.

On the other hand, the obtained confidence values strength the possibility of both the antecedent and the consequent being found together for a subject. Lift values higher than 1 tells us that the adjectives are positively correlated.

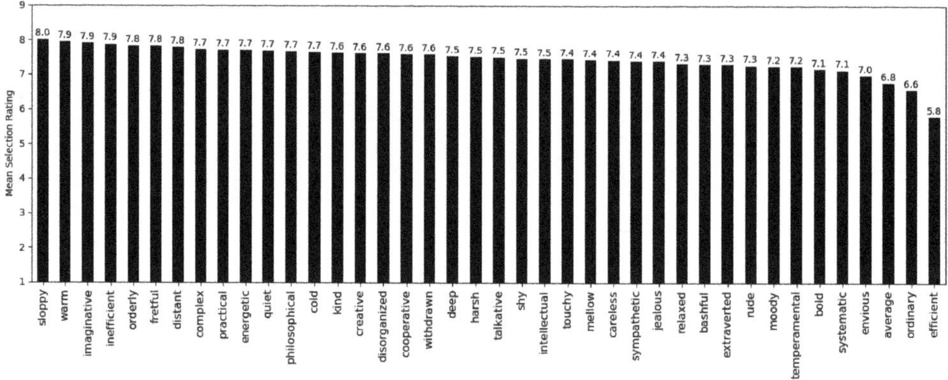

Figure 3. Mean rating values to set an adjective as selected.

Table 3. Rules with support higher than 0.15 using Association Rules Learning and the APRIORI algorithm.

Support	Confidence	Lift	Antecedent		Consequent
0.195	0.525	1.466	Creative	–>	Imaginative
0.189	0.508	1.207	Kind	–>	Imaginative
0.176	0.418	1.146	Sympathetic	–>	Kind
0.170	0.551	1.511	Sympathetic	–>	Relaxed
0.170	0.491	1.394	Withdrawn	–>	Intellectual
0.170	0.491	1.165	Kind	–>	Intellectual
0.170	0.540	1.281	Kind	–>	Shy
0.164	0.464	1.273	Sympathetic	–>	Withdrawn
0.164	0.634	1.505	Kind	–>	Jealous
0.157	0.521	1.428	Sympathetic	–>	Systematic
0.157	0.500	1.371	Sympathetic	–>	Bashful
0.157	0.500	1.187	Kind	–>	Bashful
0.151	0.510	1.400	Sympathetic	–>	Bold
0.151	0.358	1.017	Withdrawn	–>	Kind
0.151	0.585	1.389	Kind	–>	Energetic

2.1.3. Data Pre-Processing

First, a random seed, as 91,190,530, was defined for replicability purposes. Then, five observations that had abnormal values were removed. In particular, one observation had abnormally high values, while other four were filled with the same exact dummy value for all adjectives. None of these observations add the *selected_attr* feature filled. The final dataset is made of 250 observations, with the next lines describing the entire treatment and all applied methods, including synthetic data creation.

Handling Zero-Ratings.

The lowest accepted value by Saucier's test is one, however zeros are present in the dataset. To correct this situation and to make all observations mathematically valid, such values were updated to the nearest valid value, with traits' values being re-calculated based on such changes.

One-hot Encoding the Selected_attr Feature.

The *selected_attr* feature consists of a string with comma-separated adjectives. Such data was one-hot encoded using a Multi-Label Binarizer, allowing these data to become easier to handle by ML models. Forty new features were created, being entitled as *{adjective}_selected*, with *{adjective}* being a placeholder for the corresponding adjective name. A value of 0 means that the adjective was not selected, with a value of 1 meaning selection.

Filling the Selected_attr Feature When Empty.

Approximately 38% of all observations do not have adjectives selected, i.e., the *selected_attr* feature is empty because the subject did not choose any adjective. However, to be able to propose the ASAP method, we are required to have as much observations as possible with the *selected_attr* feature filled. Hence, a method was conceived to synthetically mark adjectives as selected based on adjectives' ratings and frequent patterns of selected adjectives.

The first step consists in iterating through the observations without selected adjectives. Then, for each observation, iterate through each adjective. If the adjective's rating is higher than the mean selection rating of that same adjective (as depicted in Figure 3), then the adjective is a candidate to be *selected*. Being a candidate means that the adjective may, or may not, be selected. To reduce bias, this decision is randomised, with the adjective having a three-quarters chance of being selected. If the adjective is to be selected, then the corresponding *{adjective}_selected* one-hot feature is selected (marked with 1). The next step is to see if the selected adjective is part of any rule (as depicted in Table 3). If it is, then the consequent will have half a chance of being selected as well. The upper limit is of fourteen selected adjectives per observation, with the lower limit being one selected adjective. To respect this last condition, for each observation, it is stored a list of all adjectives that are above the selection threshold. If no adjective was previously selected, than a random adjective from the referred list is selected. Algorithm 1 describes, using pseudo-code, the implemented method.

The method described in Algorithm 1 enabled all observations to have adjectives selected. Considering only the affected observations, the mean value is of 7.8 selected adjectives per observation, with a minimum of 1 and a maximum of 14 selected adjectives. Several randomized decisions are made based on a probabilistic approach in order to reduce any possible bias.

Data Augmentation.

Since the small size of the dataset may pose a problem to ML models, we aimed to investigate how models would behave on non-data scarce environments. Hence, Data Augmentation (DA) techniques were conceived to increase the dataset's size. It is worth highlighting that there is no standardised DA process that can be applied to every domain. Instead, DA refers to a process that is highly dependent of the domain where it is to be implemented. The goal is to increase the dataset size while maintaining relations and data specificities, using randomness to reduce bias.

With the use of DA techniques, a second dataset was conceived. Hence, two distinct input datasets will be fed to the candidate ML models. On the one hand, models are to be trained and evaluated with the original dataset, without any DA (*No DA*). On the other, candidate models will also be trained and evaluated using an augmented dataset (*With DA*). In the augmented dataset, new observations were generated from every single observation. The number of new observations that can be generated from one observation varies according to a random variable that outputs, with the same probability, a number between 15 and 25. For every new observation, another random variable will decide how many and which adjectives to vary from the original observation. A minimum of 5 and a maximum of 20 adjectives must vary. Each of these adjectives can stay the same or go up/down one or two units, always respecting the test limits of 1 and 9. Then, the Big Five are calculated for the new observations. Finally, the last step consists in selecting and deselecting adjectives. In particular, in finding out if the adjective that varied is a candidate to be selected or deselected, similarly to what was done to fill the *selected_attr* feature when empty. If the adjective that had its value updated is a candidate to be selected and if it was indeed chosen to be selected (three-quarters chance), then if it is an antecedent of any rule, the consequent would also have half a chance of being selected too. Finally, a final random variable, varying from 5 to 14, defines how many selected adjectives the new observation can hold. If such limit is exceeded, then, randomly, selected adjectives are deselected until the upper limit is respected.

Algorithm 1: Filling the *selected_attr* feature.

```
Input: dataset, limit = 14
adj_thresholds = getAdjectivesThresholds(dataset)
foreach row ∈ dataset do
    if row.selected_attr == 'na' then
        initialise enabled_adjectives = 0
        initialise obs_without_selection = {}
        foreach adjective ∈ row.adjectives do
            if adjective.value >= adj_thresholds[adjective] then
                if enabled_adjectives < limit then
                    if random.choice(4) < 3 then
                        enabled_adjectives += 1
                        dataset[row][{adjective}_selected] = 1
                        list_of_consequents = getConsequents(adjective)
                        foreach consequent_adjective ∈ list_of_consequents do
                            if random.choice(2) < 1 then
                                enabled_adjectives += 1
                                dataset[row][{consequent_adjective}_selected] = 1
                                if enabled_adjectives == limit then
                                    break
                                end
                            end
                        end
                    end
                end
            else
                obs_without_selection[adjective] = adjective.value
            end
        end
    end
    if enabled_adjectives == 0 then
        random_adjective = random.choice(obs_without_selection.keys())
        dataset[row][{random_adjective}_selected] = 1
    end
end
```

Data augmentation processes may add an intrinsic bias to ML models. Hence, to reduce bias to its minimum, several randomized decisions were made based on a probabilistic approach in order to create a more generalized version of the dataset.

Binning.

In Saucier's original study [13], trait scores were divided into three bins. Trait scores between [8, 29] were considered to be *low*, between [30, 50] were considered to be *average*, and between [51, 72] were considered to be *high*. This assumes an increased importance since one of the conceived ML architectures, as explained later, uses classification models, where labels (the personality traits) are required to be binned. Hence, considering the original split and the need to create trait bins, labels were binned using the three bins defined originally. As depicted in Figure 4, after binning the dataset using the original intervals, bins get imbalanced, with all five traits having a higher number of observations falling within the range [30, 50]. In fact, for all traits, around 60% of observations fall within the *average* bin. Regarding the other two bins, *high* contains significantly more observations than *low* for all traits except for the *Extroversion* trait, which contains approximately the same amount of observations in the *high* and *low* bins. This distribution of observations must be taken into consideration when conceiving and training the ML models. In fact, this distribution will lead to the use of error metrics that take into account the presence of imbalanced bins.

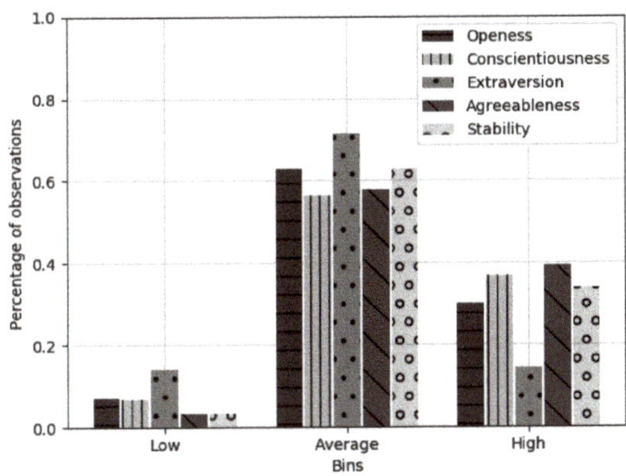

Figure 4. Distribution of observations per bin and personality trait.

Final Considerations.

Two datasets were created. *Age, language* and *genre* features were removed from both datasets as well as the rating of the 40 adjectives since those will not be used by the models. Dataset with *No DA* consists of 250 observations, while the dataset *With DA* consists of 5230 observations. Both datasets contain 50 features that correspond to the 40 one-hot encoded adjectives, the 5 personality traits' scores and the 5 binned personality traits.

2.2. Modelling

Based on the collected dataset, its characteristics and the essence of the ASAP method, two different ML architectures were conceived and evaluated. The first architecture consists of five supervised trait regressors while the second one consists of five supervised trait classifiers. The goal is to obtain the Big Five scores based on the selection of adjectives.

Both architectures use gradient boosting, in particular Gradient Boosted Trees to tackle this supervised learning problem. The "gradient boosting" term was first used by J. Friedman [28], being used as a ML technique to convert weak learners, typically Decision Trees, into strong ones, allowing the optimisation of a differentiable loss function, with the gradient representing the slope of the tangent to the loss function. Gradient boosting trains weak learners in a gradual, additive and sequential manner. A gradient descent procedure is performed so that trees are added to the gradient boosting model in order to reduce the model's loss. Being this a greedy algorithm, it can overfit. Hence, to control overfitting, it is common to use regularisation parameters, limit the number of trees of the model, and tree's depth and size. Another benefit of using Gradient Boosted Trees is the ability to compute estimates of feature importance.

2.2.1. Architecture I—Big Five Regressors

The first proposed architecture uses a total of five different Gradient Boosted Trees regression models to obtain the score of the Big Five, with each model mapping a specific trait (Figure 5). As input, each model receives the one-hot encoded adjectives' selection (whether the adjective was selected or not). The main characteristics of this architecture may be summarised as follows:

- *Input*: the one-hot encoded adjectives selection;
- *Output*: the score of each personality trait;
- *Evaluation*: two independent trials using nested cross-validation with Mean Squared Error (MSE) as objective function and Root Mean Squared Error (RMSE) and Mean Absolute Error (MAE) as evaluation metrics;
- *Models*: personality traits are computed independently of others, i.e., five independent regression models are trained, one for each trait.

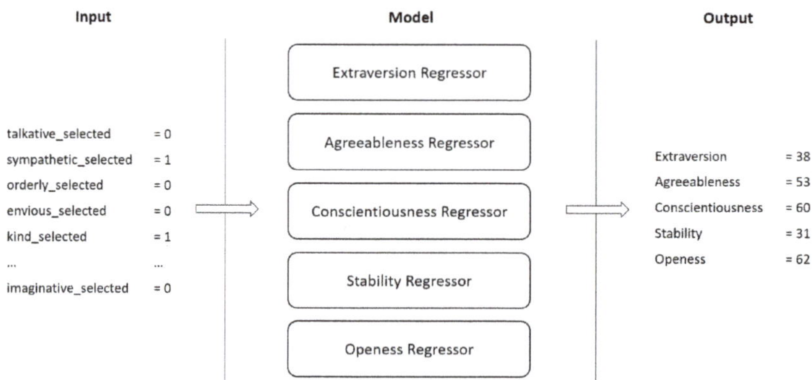

Figure 5. Architecture I—Big Five regressors.

2.2.2. Architecture Ii—Big Five Bin Classifiers

The second proposed architecture uses a total of five different Gradient Boosted Trees classification models to obtain the binned score of the Big Five, with each model mapping a specific trait (Figure 6). As input, each model receives the one-hot encoded adjectives' selection (whether the adjective was selected or not). The main characteristics of this architecture may be summarised as follows:

- *Input*: the one-hot encoded adjectives selection;
- *Output*: the bin (low/average/high) of each personality trait;
- *Evaluation*: two independent trials using nested cross-validation for multi-output multi-class classification with softmax as objective function and accuracy, f1-score and mean error as metrics;
- *Models*: personality traits are computed independently of others, i.e., five independent classification models are trained, one for each trait.

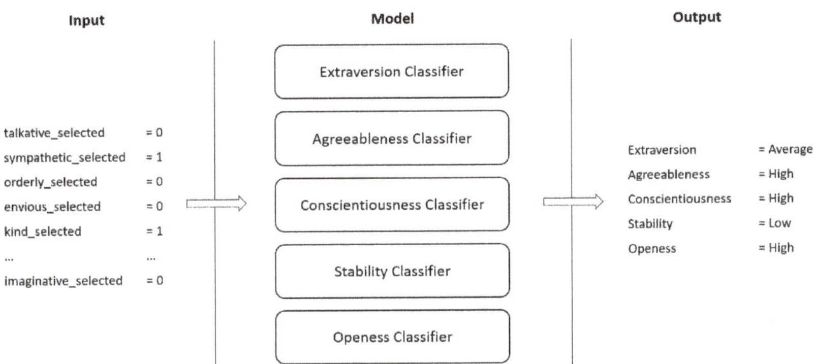

Figure 6. Architecture II—Big Five bin classifiers.

2.2.3. Models' Evaluation

All conceived models follow a Supervised Learning approach, i.e., models are trained on a sub-set of data and are then evaluated on a distinct sub-set. In fact, we went further and implemented nested cross-validation to estimate the skill of the candidate models on unseen data as well as for hyperparameter tuning. Hyperparameter selection is performed in the inner loop, while the outer one computes an unbiased estimate of the candidate's accuracy. Nested cross-validation assumes an increased importance since, otherwise, the same data would be used to tune the hyperparameters and to evaluate the model's accuracy [29]. Inner cross-validation was performed with $k = 4$ and outer cross-validation used $k = 3$. Two independent trials were performed. All candidate models were evaluated and validated against the original results from Saucier's test for each sample.

To evaluate the effectiveness of Architecture I, two error metrics were used. Both take as input the model's predicted value ($\widehat{y}$) and the actual value from Saucier's test (y), computing a metric of how far the model is from the real known value. The first one, RMSE, allows us to penalise outliers and easily interpret the obtained results since they are in the same unit of the feature that is being predicted by the model (Equation (1)). The second error metric, MAE, was used to complement and strengthen the confidence on the obtained values (Equation (2)).

$$\text{RMSE} = \sqrt{\frac{\sum_{i=1}^{n}(y_i - \widehat{y}_i)^2}{n}} \tag{1}$$

$$\text{MAE} = \frac{1}{n}\sum_{i=1}^{n}|y_i - \widehat{y}_i| \tag{2}$$

Since Architecture II consists of several classification models, confusing matrix-based metrics were used to evaluate the classifier's output quality, in particular the f1-score (Equation (3)), where the relative contribution of precision and recall are equal, and the Mean Error (Equation (4)), which penalises wrongly classified observations. Being this a multi-class problem and considering that bins are imbalanced, both micro and macro-averaged f1-scores are used. Macro-average computes the error metric independently for each class and averages the errors, treating all classes equally. On the other hand, micro-average aggregates all classes' contributions to compute the final error metric. If the goal is to maximise the models' hits and minimize its misses, micro-average should be used since it aggregates the results of all classes before computing the final error metric. On the other hand, if the minority classes are more important, a macro-averaged approach would be useful since it is insensitive to the imbalance of the classes by computing the error metric independently for each class and then averaging all errors from all classes.

$$\text{F1 score} = 2 \times \frac{precision \times recall}{precision + recall} \tag{3}$$

$$ME = \frac{\text{Wrongly Classified Observations}}{\text{Total Number of Observations}} \tag{4}$$

2.3. Experiments

The conceived architectures focus on quantifying the Big Five of a subject based on a selection of adjectives that describe him/her the most. For both architectures, experiments were conducted under the same settings and conditions. The same random seed was used.

2.3.1. Experimental Setup

Python, version 3.7, was the used programming language for data exploration and pre-processing as well as for model development and evaluation. *Pandas, NumPy, scikit-learn, XGBoost, matplotlib* and *seaborn* were the used libraries. The *Knime* platform was also used for data exploration. All hardware was made available by Google's Colaboratory, a free python environment that requires minimal setup and runs entirely in the cloud.

XGBoost was the library used to conceive the Gradient Boosted Trees. It is a distributed gradient boosting library that is efficient and flexible. Contrary to other boosted trees based libraries, XGBoost implements regularisation and parallel processing, having already been used in several studies [30–32]. Algorithm 2 describes, using pseudo-code, the method used to conceive the boosted regressors and classifiers, depending on the inputted architecture.

Algorithm 2: Building the Gradient Boosted models.

Input: *architecture*
if *architecture* == *1* **then**
    estimator = XGBRegressor(booster = 'gbtree', objective = 'reg:squarederror')
    multi_estimator = MultiOutputRegressor(estimator)
else
    estimator = XGBRegressor(booster = 'gbtree', objective = 'multi:softmax', num_class = 3)
    multi_estimator = MultiOutputClassifier(estimator)
end
return multi_estimator

2.3.2. Hyperparameter Search Space

Models were tuned in regard to a set of hyperparameters using Random Search limited to 175 combinations (out of 486). Architecture I uses MSE as objective function while Architecture II uses softmax. Table 4 describes the searching space for each hyperparameter.

Table 4. Models hyperparameters' searching space.

Parameter	Searched Values	Rationale
a. Number of Estimators	[300, 400, 500]	Number of trees in a model
b. Eta	[0.01, 0.05, 0.1]	Learning rate
c. Gamma	[0.02, 0.04, 0.08]	Minimum loss reduction required to make a further partition on a leaf node
d. Trees' Max Depth	[4, 12, 18]	Maximum depth of a tree
e. Minimum Child Weight	[4, 6, 8]	Minimum sum of instance weight needed in a child (higher values for more conservative models)
f. Colsample by tree	[0.2, 0.3]	Fraction of columns to be sub-sampled (controlling correlation between trees)

3. Results

Two distinct ML architectures were experimented. One uses Gradient Boosted Trees regressors to obtain the exact value of each personality trait (Architecture I) while the other uses Gradient Boosted Trees classifiers to obtain the bin of each personality trait (Architecture II). Different experiments were conducted with two distinct datasets. One with 250 observations (*No DA*) and another with 5230 observations (*With DA*). Both architectures receive, as input, the one-hot encoded selection of adjectives.

Nested cross-validation was performed to tune the hyperparameters and to have a stronger validation of the obtained results. Inner cross-validation was performed using $k = 4$, with random search being used to find the best set of hyperparameters. In the inner loop, 700 fits were performed (4 folds $\times$ 175 combinations). The outer cross-validation loop used $k = 3$, totalling 2100 fits (3 folds $\times$ 700 fits). Two independent training trials were performed, with a grand total of 4200 fits (2 trials $\times$ 2100 fits) per architecture per dataset.

3.1. Architecture I—Big Five Regressors

All candidate models were evaluated in regard to RMSE and MAE error metrics. Table 5 depicts the best hyperparameter configuration for Architecture I, for both datasets. What immediately stands out is the better performance of the candidate models when using the larger dataset. In fact, RMSE decreases about 30% when using the dataset *With DA*. This was already expected since the dataset with *No DA* was made of only 250 observations.

Overall, for Architecture I with *No DA* the error is of approximately 8 units of measure. Since RMSE outputs an error in the same unit of the features that are being predicted by the model, it means that this Architecture is able to obtain the value of each personality trait with an error of 8 units. On the other hand, for Architecture I *With DA*, RMSE is of approximately 5.6 units of measure. It is also possible to discern that RMSE tends to be more stable when using the *With DA* dataset when compared to the *No DA* dataset which shows higher error variance. In Table 5, the *Evaluation* column presents the error value of the best candidate model in the outer test fold. These values provide a second and stronger validation of the ability to classify of the best model per split.

Table 5. Architecture I results with and without data augmentation, for each independent trial, with RMSE as metric. Hyperparameters described by letters as follows: *a.* number of estimators, *b.* eta, *c.* gamma, *d.* trees' max depth, *e.* minimum child weight and *f.* colsample by tree.

Trial	CV Split	Best Score	Evaluation	Fit Time (min)	*a.*	*b.*	*c.*	*d.*	*e.*	*f.*
				No Data Augmentation						
1	1	7.813	8.078	3.8	300	0.05	0.04	4	6	0.2
1	2	8.015	7.560	3.8	300	0.05	0.02	4	4	0.2
1	3	8.203	7.512	3.7	300	0.01	0.04	4	4	0.2
2	1	8.024	7.594	3.7	300	0.10	0.08	4	8	0.2
2	2	8.184	7.161	3.7	300	0.05	0.02	4	8	0.2
2	3	7.847	7.961	3.7	300	0.05	0.04	4	8	0.2
				With Data Augmentation						
1	1	5.692	5.464	64.8	300	0.10	0.02	12	4	0.3
1	2	5.604	5.602	65.9	300	0.01	0.02	18	4	0.3
1	3	5.646	5.520	69.6	300	0.01	0.02	18	4	0.3
2	1	5.637	5.537	68.4	300	0.01	0.02	12	4	0.3
2	2	5.632	5.482	65.7	300	0.01	0.04	18	6	0.3
2	3	5.673	5.467	67.4	300	0.01	0.08	12	4	0.3

The hyperparameter tuning process is significantly faster for Architecture I with *No DA*, taking around 3.7 min to perform 700 fits and around 22 min to perform the full run. On the other hand,

Architecture I *With DA* takes more than 1 hour to perform the same amount of fits, requiring more than 6.5 hours to complete. Overall, the models that behaved the best used 300 gradient boosted trees. Interestingly, when using the dataset with *No DA*, all models required 20% of the entire feature set when constructing each tree (colsample by tree) and used a maximum depth of 4 levels, building shallower trees which helps controlling overfitting in the smaller dataset. On the other hand, when using the dataset *With DA*, the best models not only required 30% of the feature set but also required deeper trees, which indicate the need for more complex trees to find relations in the larger dataset. To strengthen this assertion, the learning rate is also smaller in Architecture I *With DA* allowing models to move slower through the gradient.

Focusing the results obtained from testing in the test fold of the outer-split, Architecture I *With DA* presents a global RMSE of 5.512 and MAE of 3.979. On the other hand, Architecture I with *No DA* presents higher error values, with a global RMSE and MAE of 7.644 and 6.082, respectively. The fact that RMSE and MAE have relatively close values implies that not many outliers, or distant classifications, were provided by the models. It is also interesting to note that, independently of the dataset, *Openness* is the most difficult trait to classify. All these data is given by Table 6, where the MSE is also displayed, being used to compute the RMSE.

Table 6. Evaluation results of Architecture I, with and without data augmentation, obtained from the test folds of the outer-split.

Metric	Global	Extraversion	Agreeableness	Conscient.	Stability	Openness
No Data Augmentation						
MAE	**6.082**	5.495	5.942	5.616	6.205	7.153
MSE	**58.527**	45.973	55.984	49.230	58.933	82.514
RMSE	**7.644**	6.778	7.468	7.000	7.668	9.061
With Data Augmentation						
MAE	**3.979**	3.937	4.071	3.832	3.798	4.259
MSE	**30.385**	29.529	31.630	28.813	27.069	34.884
RMSE	**5.512**	5.433	5.623	5.367	5.201	5.906

Figure 7 provides a graphical view of RMSE and MAE for Architecture I for both datasets, being possible to discern that both metrics present a lower error value when conceiving models over the augmented dataset.

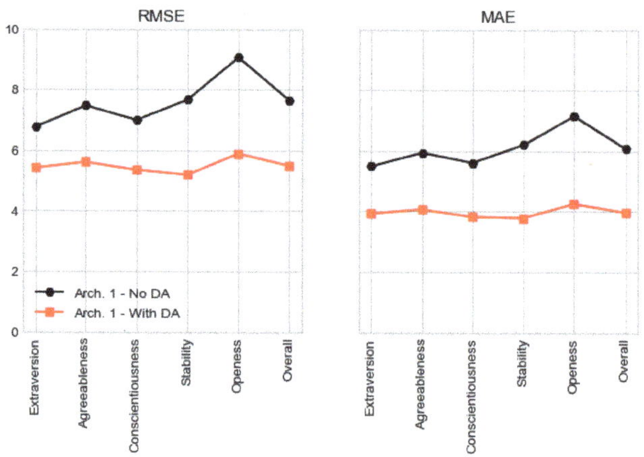

Figure 7. Graphical view of Architecture's I RMSE and MAE for both datasets.

3.2. Architecture Ii—Big Five Bin Classifiers

Architecture II candidate models, which classify personality traits in three bins (low, average and high), were evaluated using several classification metrics. Table 7 depicts the best hyperparameter configuration for Architecture II, for the two datasets, using accuracy as metric. Again, models conceived over the dataset *With DA* outperform those conceived over the dataset with *No DA*, more than doubling the accuracy value. In addition, their evaluation values also tend to be more stable and less prone to variations. However, one may argue that the accuracy values attained by the candidate models and presented in Table 7 are low. Hence, it is of the utmost importance to assert that such accuracy values correspond to samples that had all five traits correctly classified. I.e., if one trait of a sample was wrongly classified, than that sample would be considered as badly-classified even if the remaining four traits were correctly classified. To provide a stronger validation metric, Table 8 provides metrics based on traits' accuracy instead of samples' accuracy, presenting significantly higher values.

Table 7. Architecture II results with and without data augmentation, for each independent trial, with sample accuracy as metric. Hyperparameters described by letters as follows: *a.* number of estimators, *b.* eta, *c.* gamma, *d.* trees' max depth, *e.* minimum child weight and *f.* colsample by tree.

Trial	CV Split	Best Score	Evaluation	Fit Time (min)	*a.*	*b.*	*c.*	*d.*	*e.*	*f.*
				No Data Augmentation						
1	1	0.144	0.131	8.4	500	0.05	0.02	4	6	0.3
1	2	0.180	0.181	8.6	400	0.01	0.08	4	4	0.2
1	3	0.138	0.108	8.4	300	0.01	0.04	12	8	0.3
2	1	0.175	0.143	8.5	300	0.05	0.04	4	4	0.2
2	2	0.138	0.181	8.4	500	0.10	0.02	18	4	0.3
2	3	0.155	0.133	8.4	400	0.05	0.08	18	8	0.3
				With Data Augmentation						
1	1	0.458	0.486	128.6	300	0.10	0.02	12	4	0.3
1	2	0.464	0.466	130.1	300	0.01	0.08	12	4	0.3
1	3	0.453	0.468	133.6	400	0.10	0.08	12	4	0.2
2	1	0.465	0.490	129.4	300	0.10	0.02	18	4	0.3
2	2	0.466	0.466	123.6	300	0.01	0.04	12	4	0.3
2	3	0.448	0.480	125.3	500	0.10	0.08	18	4	0.2

Still regarding Table 7, it becomes clear that the tuning process is significantly faster for Architecture II with *No DA*, taking around 50 min to complete the process. On the other hand, when using the larger dataset, the process takes more than 12 hours to complete. Overall, models tend to use 300 gradient boosted trees and require 30% of the entire feature set per tree. The best classifiers also require deeper trees, with 12 or 18 levels. It is also worth mentioning that all the best models conceived over the dataset *With DA* required a minimum child weight of 4. This hyperparameter defines the minimum sum of weights of all observations required in a child node, being used to control overfitting and prevent under-fitting, which may happen if high values are used when setting this hyperparameter.

As stated previously, all metrics provided in Table 8 are based on traits' accuracy. Using class accuracy instead of sample accuracy, the mean error of Architecture II candidate models using the dataset *With DA* is of 0.165, which corresponds to an accuracy higher than 83%. On the other hand, the mean error with *No DA* increases to 0.338. Overall, all models show better results when using the dataset *With DA*.

In this study, both micro and macro-averaged metrics were evaluated. However, since we are interested in maximising the number of correct predictions each classifier makes, special importance is given to micro-averaging. In fact, micro f1-score of the classifiers conceived over the dataset *With DA* display an interesting overall value of 0.835, with the *Openness* trait being, again, the one showing the lower value. It is worth mentioning that micro-averaging in a multi-class setting with all labels included, produces the same exact value for the f1-score, precision and recall metrics, being this the reason why Table 8 only displays micro f1-score. On the other hand, macro-averaging computes each error metric independently for each class and then averages the metrics, treating all classes equally. Hence, since models depict a lower macro f1-score when compared to the micro one, this could mean that there may be some classes that are less used when classifying, such as *low* or *high*. Nonetheless, macro f1-score still present a very interesting global value of 0.776. Macro-averaged precision also depicts a high value, strengthening the ability of models to correctly classify true positives and avoid false positives. Finally, models' global macro-averaged recall is of 0.742, still a significant value that tells us that the best candidate models are able, in some extent, to avoid false negatives.

Table 8. Evaluation results of Architecture II, with and without data augmentation, based on trait's accuracy and obtained from the test folds of the outer-split.

Metric	Global	Extraversion	Agreeableness	Conscient.	Stability	Openness
No Data Augmentation						
Mean Error	**0.338**	-	-	-	-	-
Micro F1-Score	**0.663**	0.728	0.660	0.620	0.648	0.656
Macro F1-Score	**0.459**	0.532	0.462	0.419	0.443	0.438
Macro Precision	**0.477**	0.590	0.468	0.413	0.445	0.468
Macro Recall	**0.464**	0.525	0.469	0.434	0.447	0.447
With Data Augmentation						
Mean Error	**0.165**	-	-	-	-	-
Micro F1-Score	**0.835**	0.846	0.834	0.831	0.843	0.822
Macro F1-Score	**0.776**	0.770	0.795	0.801	0.731	0.782
Macro Precision	**0.830**	0.826	0.854	0.840	0.809	0.819
Macro Recall	**0.742**	0.731	0.758	0.774	0.691	0.755

Figure 8 provides a graphical view of micro and macro-averaged f1-score and precision for Architecture II for both datasets, being again possible to recognise a better performance when using the dataset *With DA*.

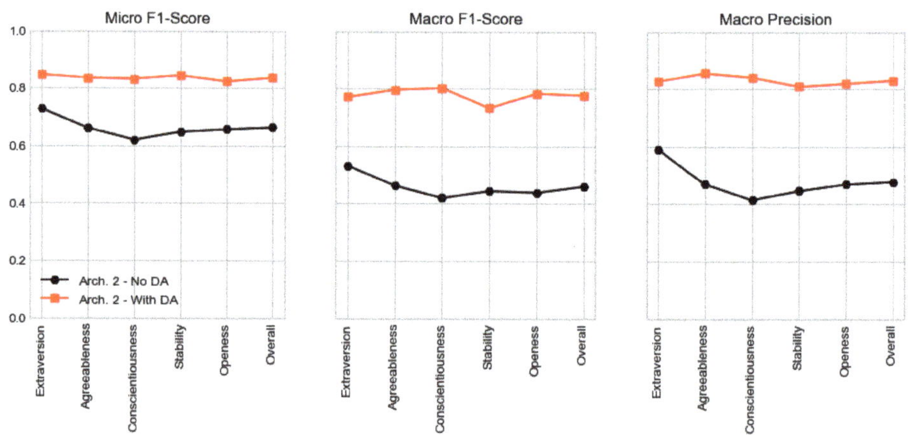

Figure 8. Graphical view of Architecture's II micro and macro-averaged f1-score and precision for both datasets.

3.3. Feature Importance

Gradient Boosted Trees allow the possibility of estimating feature importance, i.e., a score that measures how useful each feature was when building the boosted trees. This importance was estimated using *gain* as importance type, which corresponds to the improvement in accuracy brought by a feature to the branches it is on. A higher value for a feature when compared to another, implies it is more important for classifying the label.

Figure 9 presents the estimated feature importance of Architecture I using an heat-map view. Interestingly, models conceived using the dataset with *No DA* (Figure 9a) give an higher importance to the selection of the adjective *inefficient* when classifying the *Conscientiousness* trait. *Sloppy, disorganized* and *careless* are other adjectives that assume special relevance when classifying the same personality trait. Regarding the *Extraversion* trait, *talkative, quiet* and *withdrawn* are the most important adjectives, being only then followed by the *extroverted* and *energetic* ones. The *Agreeableness* trait gives higher importance to *distant, harsh, cold* and *rude*. On the other hand, feature importance is more uniform in the *Stability* and *Openness* personality traits, with the most important adjectives assuming a relative importance of about 7%. Another interesting fact that arises from these results, is that some adjectives have lower importance for all five traits. Examples include *bashful, bold, intellectual* and *jealous*.

As for the models conceived using the dataset *With DA* (Figure 9b), results are similar to the smaller dataset. In these models there are less important features, but the ones considered as important have a stronger importance. An example is the case of the adjective *talkative* for the *Extraversion* trait, which increases its importance from 16% to 22%, and quiet, which increases from 11% to 17%. *Withdrawn* and *quiet* have a reduced importance. Interestingly, for the *Agreeableness* trait, the adjective *kind* becomes the most important one, increasing from 3.2% to 15%. The *Openness* trait still assumes a more uniform importance for all features, being this one of the reasons why it was the trait showing worst performance using Architecture I models.

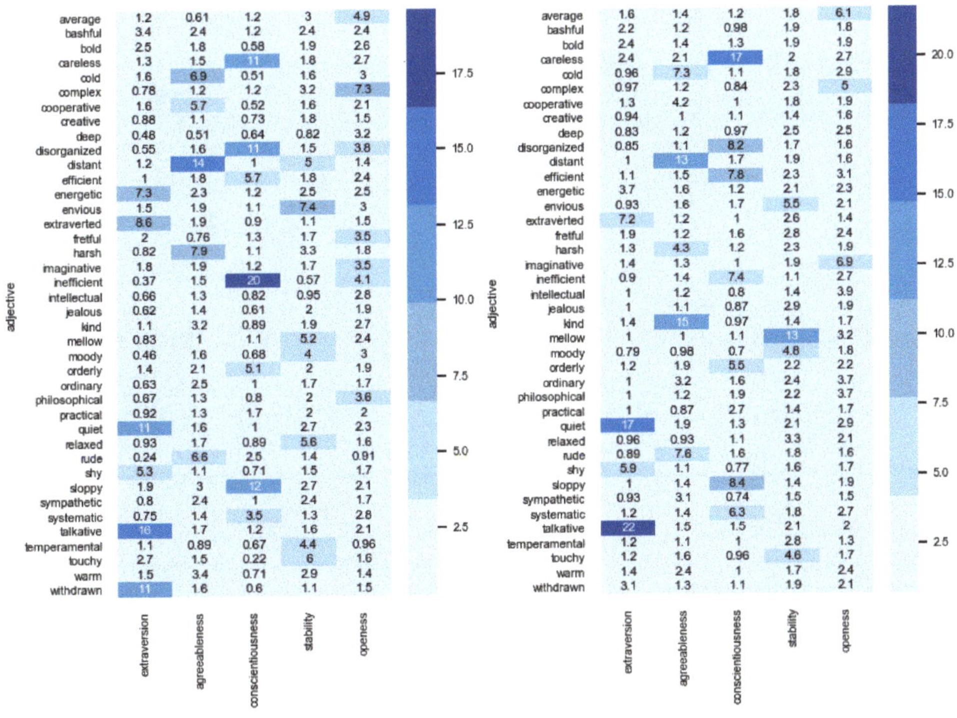

(a) Using dataset with *No DA*. (b) Using dataset *With DA*.

Figure 9. Feature importance heat-map of Architecture I.

Regarding Architecture II, Figure 10 presents the estimated feature importance for both datasets. What immediately draws one attention is the fact that importance values are much more balanced when compared to Architecture I. Indeed, the highest importance value is of 9.1% with *No DA* and 13% *With DA* when compared to 20% and 22% of Architecture I, respectively. Nonetheless, except for a few exceptions, adjectives assuming higher importance in Architecture I also assume higher importance in Architecture II. The main difference is that values are closer together, having a lower amplitude.

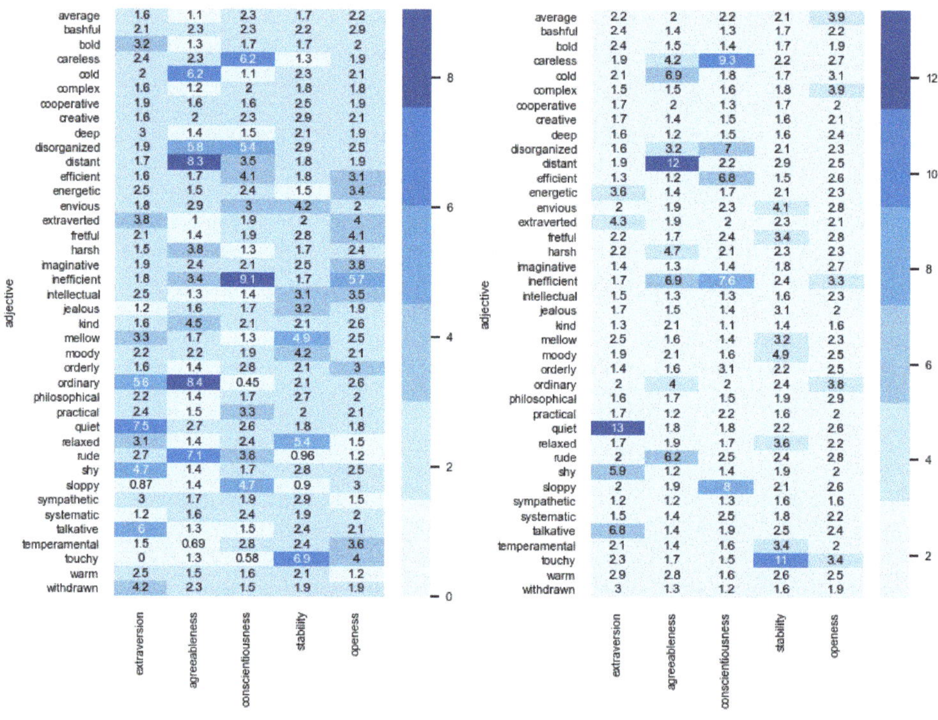

(a) Using dataset with *No DA*.   (b) Using dataset *With DA*.

Figure 10. Feature importance heat-map of Architecture II.

4. Discussion and Conclusions

The proposed ASAP method aims to use ML-based models to reinstate the process of rating adjectives or answering questions by an adjective selection process. To achieve this goal, two different ML architectures were proposed, experimented and evaluated. The first architecture uses Gradient Boosted Trees regressors to quantify the Big Five personality traits. Overall, this architecture is able to quantify such traits with an error of approximately 5.5 units of measure, providing an accurate output given the limited amount of available records. On the other hand, Architecture II uses Gradient Boosted Trees classifiers to qualify the bin in which the subject stands, for each trait. Bins are based on Saucier's original study where trait scores between [8, 29] are considered *Low*, between [30, 50] are considered *Average*, and between [51, 72] are considered *High*. This architecture was able to quantify the personality traits with a micro-averaged f1-score of more than 83%. A better performance of both architectures in the augmented dataset was also expected since the original dataset had a limited amount of records. The implemented data augmentation techniques aimed to increase the dataset size following well-defined rationales but also included several randomised decisions based on a probabilistic approach in order to reduce bias and create a more generalised version of the dataset. For this, data exploration and pattern mining, in the form of Association Rules Learning, assumed an increased importance, allowing us to understand relations between selected adjectives. Results for records with very few adjectives selected may be biased to the dataset used to train the models since the ability to quantify traits based on the selection of just one or two adjectives is of an extreme difficulty. Hence, for the ASAP method to behave properly, subjects should be encouraged to select four, or more, adjectives.

A further validation was carried out by means of a significance analysis between the correlation differences of predicted and actual scores. The best overall candidate model of Architecture I was trained using, as input data, 90% of the original dataset, with the remaining being used to obtain predictions. Predictions were compared with the actual scores of the five traits. As expected, the p-value returned an high value (0.968), with a z-score of 0.039. Such values tell, with a high degree of confidence, that the null hypothesis should be retained and that both correlation coefficients are not significantly different from each other. This is in line with expectations since the conceived models are optimizing a differentiable loss function, using a gradient descent procedure that reduces the model's loss to increase the correlation between predictions and actual scores.

Architecture II took significantly more time to fit than Architecture I. However, it provides more accurate results, which are less prone to error. It should be noted that Architecture II only provides an approximation to the Big Five of the subject, i.e., it does not numerically quantify each trait, instead it tells in which bin the subject finds himself. This can be useful in cases where the general qualification of each trait is more important than the specific score of the trait. On the other hand, Architecture I will provide an exact score for each personality trait based on a selection of adjectives. Indeed, the working hypothesis has been confirmed, i.e., it is possible to achieve promising performances using ML-based models where the subject, instead of rating forty adjectives or answering long questions, selects the adjectives he relates the most with. This allows one to obtain the Big Five using a method with a reduced complexity and that takes a small amount of time to complete. Obviously, the obtained results are just estimates, with an underlying error. The conducted experiments shown the ability of ML-based models to compute estimates of personality traits, and should not be seen as a definitive psychological assessment of one's personality traits. For a full personality assessment, tests such as the one proposed by Saucier, Goldberg or the NEO-personality-inventory should be used.

The use of augmented sets of data may bring an intrinsic bias to the candidate models. In all cases, preference should always be given to the collection and use of real data. However, in scenarios where data is extremely costly, an approximation may allow ML models to be analyzed with augmented data. In such scenarios, data augmentation processes should make use of several randomized decisions based on probabilistic approaches to create a generalized version of the smaller dataset. Experiments should be carefully conducted, implementing two, or more, independent trials, cross-validation and even nested cross-validation. Models, when deployed, should monitor their performance and, in situations with a clear performance degradation, should be re-trained with new collected data.

In Saucier's test, each personality trait is computed using the rating of eight unipolar adjectives, i.e, no adjective is used for more than one personality trait. Indeed, it is known, beforehand, which adjectives are used by each trait. For example, the *Extroversion* trait is computed based on four positively weighted adjectives (*extroverted, talkative, energetic* and *bold*) and four negative ones (*shy, quiet, withdrawn* and *bashful*). However, in the proposed ML architectures that make the ASAP method, all 40 adjectives are used to compute all traits, allowing the ML models to use adjectives selection/non-selection to compute several traits, thus harnessing inter-trait relationships. For instance, *bold*, one of the adjectives used by Saucier to compute *Extroversion*, shows a small importance in the conceived architectures when quantifying *Extroversion*. The same happens for *bashful* in *Extroversion*, *creative* in *Openness*, and *practical* in *Conscientiousness*, just to point a few. This could lead us to hypothesise that, one, the list of forty adjectives could be further reduced to a smaller set of adjectives by removing those that are shown to have a smaller importance and that, two, there are adjectives that can be used to quantify distinct personality traits, such as the case of *disorganised*, which can be used for the *Conscientiousness* and the *Agreeableness* traits. It is also interesting to note the lack of features assuming high importance when quantifying *Openness*. In fact, one of its adjectives, *ordinary*, seems to assume higher importance in the *Agreeableness* trait. Overall, Saucier's adjective-trait relations are being found and used by the conceived models.

Since the conceived ML architectures proved to be both performant and efficient using a selection of adjectives, future research points towards a reduction to the minimum required set of adjectives that does not harm the method's accuracy, further reducing complexity and the time it takes to be performed by the subject.

Author Contributions: Conceptualization, B.F., M.C. and C.A.; methodology, B.F. and M.C.; software, B.F. and M.C.; validation, B.F. and A.G.-B.; formal analysis, B.F. and A.G.-B.; investigation, B.F. and M.C.; resources, P.N., J.N. and C.A.; data curation, B.F. and A.B.; writing–original draft preparation, B.F. and J.N.; writing–review and editing, P.N. and C.A.; supervision, C.A. and J.N.; project administration, P.N., J.N. and C.A.; funding acquisition, B.F., P.N., J.N. and C.A. All authors have read and agreed to the published version of the manuscript.

Funding: This work has been supported by FCT - *Fundação para a Ciência e a Tecnologia* within the R&D Units Project Scope: UIDB/00319/2020. It was also partially supported by a Portuguese doctoral grant, SFRH/BD/130125/2017, issued by FCT in Portugal.

Conflicts of Interest: The authors declare no conflict of interest.

Abbreviations

The following abbreviations are used in this manuscript:

ANN	Artificial Neural Networks
ARL	Association Rules Learning
ASAP	Adjective Selection to Assess Personality Test
AUC	Area Under the Curve
CART	Classification and Regression Trees
DA	Data Augmentation
DL	Deep Learning
MAE	Mean Absolute Error
ML	Machine Learning
MSE	Mean Squared Error
NEO-PI-R	NEO-personality-inventory
RMSE	Root Mean Squared Error
TIPI	Ten Item Personality Inventory

References

1. Barlett, C.; Anderson, C. Direct and indirect relations between the Big 5 personality traits and aggressive and violent behavior. *Personal. Individ. Differ.* **2012**, *52*, 870–875, doi:10.1016/j.paid.2012.01.029. [CrossRef]
2. Rothmann, S.; Coetzer, E.P. The big five personality dimensions and job performance. *J. Ind. Psychol.* **2003**, *29*, doi:10.4102/sajip.v29i1.88. [CrossRef]
3. Orzeck, T.; Lung, E. Big-five personality differences of cheaters and non-cheaters. *Curr. Psychol.* **2005**, *24*, 274–286, doi:10.1007/s12144-005-1028-3. [CrossRef]
4. Kazdin, A. *Encyclopedia of Psychology*; American Psychological Association: Washington, DC, USA, 2000; Volume 3.
5. Salgado, J. Big Five Personality Dimensions and Job Performance in Army and Civil Occupations: A European Perspective. *Hum. Perform.* **1998**, *11*, 271–288, doi:10.1080/08959285.1998.9668034. [CrossRef]
6. Ashton, M.; Lee, K. The HEXACO–60: A Short Measure of the Major Dimensions of Personality. *J. Personal. Assess.* **2009**, *91*, 340–345, doi:10.1080/00223890902935878. [CrossRef] [PubMed]
7. Myers, I. *The Myers-Briggs Type Indicator: Manual (1962)*; Consulting Psychologists Press: Palo Alto, CA, USA, 1962; doi:10.1037/14404-000. [CrossRef]
8. Riso, D.; Hudson, R. *Understanding the Enneagram: The Practical Guide to Personality Types*; Houghton Mifflin Harcourt: Boston, MA, USA, 2000.
9. Costa, P., Jr.; McCrae, R. Domains and facets: Hierarchical personality assessment using the Revised NEO Personality Inventory. *J. Personal. Assess.* **1995**, *64*, 21–50, doi:10.1207/s15327752jpa6401_2. [CrossRef] [PubMed]

10. Goldberg, R. The development of markers for the Big-Five factor structure. *Psychol. Assess.* **1992**, *4*, 26, doi:10.1037/1040-3590.4.1.26. [CrossRef]

11. John, O.; Srivastava, S. The Big Five trait taxonomy: History, measurement, and theoretical perspectives. In *Handbook of Personality: Theory and Research*; Guilford Publications: New York, NY, USA, 1999; Volume 2, pp. 102–138.

12. Kosinski, M.; Bachrach, Y.; Kohli, P.; Stillwell, D.; Graepel, T. Manifestations of user personality in website choice and behaviour on online social networks. *Mach. Learn.* **2014**, *95*, 357–380, doi:10.1007/s10994-013-5415-y. [CrossRef]

13. Saucier, G. Mini-Markers: A brief version of Goldberg's unipolar Big-Five markers. *J. Personal. Assess.* **1994**, *63*, 506–516, doi:10.1207/s15327752jpa6303_8. [CrossRef] [PubMed]

14. Majumder, N.; Poria, S.; Gelbukh, A.; Cambria, E. Deep Learning-Based Document Modeling for Personality Detection from Text. *IEEE Intell. Syst.* **2017**, *32*, 74–79, doi:10.1109/MIS.2017.23. [CrossRef]

15. Yu, J.; Markov, K. Deep learning based personality recognition from Facebook status updates. In Proceedings of the IEEE 8th International Conference on Awareness Science and Technology (iCAST), Taichung, Taiwan, 8 November 2017; pp. 383–387, doi:10.1109/ICAwST.2017.8256484. [CrossRef]

16. Sumner, C.; Byers, A.; Boochever, R.; Park, G. Predicting Dark Triad Personality Traits from Twitter Usage and a Linguistic Analysis of Tweets. In Proceedings of the 11th International Conference on Machine Learning and Applications, Boca Raton, FL, USA, 12–15 December 2012; Volume 2, pp. 386–393, doi:10.1109/ICMLA.2012.218. [CrossRef]

17. Pennebaker, J.; King, A. Linguistic styles: Language use as an individual difference. *J. Personal. Soc. Psychol.* **1999**, *77*, 1296–1312, doi:10.1037/0022-3514.77.6.1296. [CrossRef]

18. Park, G.; Schwartz, H.; Eichstaedt, J.; Kern, M.; Kosinski, M.; Stillwell, D.; Ungar, L.; Seligman, M. Automatic personality assessment through social media language. *J. Personal. Soc. Psychol.* **2015**, *108*, 934–952, doi:10.1037/pspp0000020. [CrossRef] [PubMed]

19. Schwartz, H.; Eichstaedt, J.; Kern, M.; Dziurzynski, L.; Ramones, S.; Agrawal, M.; Shah, A.; Kosinski, M.; Stillwell, D.; Seligman, M.; et al. Personality, Gender, and Age in the Language of Social Media: The Open-Vocabulary Approach. *PLoS ONE* **2013**, *8*, doi:10.1371/journal.pone.0073791. [CrossRef] [PubMed]

20. Cerasa, A.; Lofaro, D.; Cavedini, P.; Martino, I.; Bruni, A.; Sarica, A.; Mauro, D.; Merante, G.; Rossomanno, I.; Rizzuto, M.; et al. Personality biomarkers of pathological gambling: A machine learning study. *J. Neurosci. Methods* **2018**, *294*, 7–14, doi:10.1016/j.jneumeth.2017.10.023. [CrossRef] [PubMed]

21. Mehta, Y.; Majumder, N.; Gelbukh, A.; Cambria, E. Recent trends in deep learning based personality detection. *Artif. Intell. Rev.* **2019**, doi:10.1007/s10462-019-09770-z. [CrossRef]

22. Levitan, S.; Levitan, Y.; An, G.; Levine, M.; Levitan, R.; Rosenberg, A.; Hirschberg, J. Identifying Individual Differences in Gender, Ethnicity, and Personality from Dialogue for Deception Detection. In Proceedings of the Second Workshop on Computational Approaches to Deception Detection, San Diego, CA, USA, 17 June 2016; pp. 40–44, doi:10.18653/v1/W16-0806. [CrossRef]

23. Levitan, S.; Levine, M.; Hirschberg, J.; Cestero, N.; An, G.; Rosenberg, A. Individual Differences in Deception and Deception Detection. Available online: Https://Www.Semanticscholar.Org/Paper/Individual-Differences-In-Deception-And-Deception-Levitan-Levine/295332ebfb77387f4ccbacbd214edf72caf3e331 (accessed on 3 March 2020).

24. Gurpinar, F.; Kaya, H.; Salah, A. Combining Deep Facial and Ambient Features for First Impression Estimation. In Proceedings of the Computer Vision—ECCV 2016 Workshops, Amsterdam, The Netherlands, 11–14 October 2016; Volume 9915, doi:10.1007/978-3-319-49409-8_30. [CrossRef]

25. Güçlütürk, Y.; Güçlü, U.; Pérez, M.; Escalante, H.; Baró, X.; Andujar, C.; Guyon, I.; Junior, J.; Madadi, M.; Escalera, S.; et al. Visualizing Apparent Personality Analysis with Deep Residual Networks. In Proceedings of the IEEE International Conference on Computer Vision Workshops (ICCVW), Venice, Italy, 22–29 October 2017; pp. 3101–3109, doi:10.1109/ICCVW.2017.367. [CrossRef]

26. Zhang, C.; Zhang, H.; Wei, X.; Wu, J. Deep Bimodal Regression for Apparent Personality Analysis. In Proceedings of the Computer Vision—ECCV 2016 Workshops, Amsterdam, The Netherlands, 11–14 October 2016; Volume 9915, doi:10.1007/978-3-319-49409-8_25. [CrossRef]

27. Wessa, P. Cronbach alpha (v1.0.5) in Free Statistics Software (v1.2.1). Available online: https://www.wessa.net/rwasp_cronbach.wasp/ (accessed on 1 May 2020).

28. Friedman, J. Greedy Function Approximation: A Gradient Boosting Machine. *Ann. Stat.* **2001**, *29*, 1189–1232. [CrossRef]
29. Cawley, G.; Talbot, N. On Over-fitting in Model Selection and Subsequent Selection Bias in Performance Evaluation. *J. Mach. Learn. Res.* **2010**, *11*, 2079–2107.
30. Yu, W.; Na, Z.; Fengxia, Y.; Yanping, G. Magnetic resonance imaging study of gray matter in schizophrenia based on XGBoost. *J. Integr. Neurosci.* **2018**, *17*, 331–336, doi:10.31083/j.jin.2018.04.0410. [CrossRef]
31. Sahoo, D.; Balabantaray, R. Single-Sentence Compression using XGBoost. *Int. J. Inf. Retr. Res.* **2019**, *9*, 11, doi:10.4018/IJIRR.2019070101. [CrossRef]
32. Pesantez-Narvaez, J.; Guillen, M.; Alcañiz, M. Predicting Motor Insurance Claims Using Telematics Data — XGBoost versus Logistic Regression. *Risks* **2019**, *7*, 16, doi:doi.org/10.3390/risks7020070. [CrossRef]

Article

Bioinspired Hybrid Model to Predict the Hydrogen Inlet Fuel Cell Flow Change of an Energy Storage System

Héctor Alaiz-Moretón [1,†], Esteban Jove [2,†], José-Luis Casteleiro-Roca [2,†], Héctor Quintián [2,*,†], Hilario López García [3,†], José Alberto Benítez-Andrades [1,†], Paulo Novais [4,†] and Jose Luis Calvo-Rolle [2,†]

[1]   Department of Electrical and Systems Engineering, University of León, 24071 León, Spain; hector.moreton@unileon.es (H.A.-M.); jbena@unileon.es (J.A.B.-A.)

[2]   Department of Industrial Engineering, University of A Coruña, 15405 Ferrol, Spain; esteban.jove@udc.es (E.J.); jose.luis.casteleiro@udc.es (J.-L.C.-R.); jlcalvo@udc.es (J.L.C.-R.)

[3]   Department of Electrical, Electronic, Computers and Systems Engineering, University of Oviedo, 33204 Gijón, Spain; hilario@uniovi.es

[4]   Department of Informatics/Algoritmi Center, University of Minho, 4710-057 Braga, Portugal; pjon@di.uminho.pt

*   Correspondence: hector.quintian@udc.es; Tel.: +34-881-013-117

†   These authors contributed equally to this work.

Received: 16 October 2019; Accepted: 5 November 2019; Published: 7 November 2019

Abstract: The present research work deals with prediction of hydrogen consumption of a fuel cell in an energy storage system. Due to the fact that these kind of systems have a very nonlinear behaviour, the use of traditional techniques based on parametric models and other more sophisticated techniques such as soft computing methods, seems not to be accurate enough to generate good models of the system under study. Due to that, a hybrid intelligent system, based on clustering and regression techniques, has been developed and implemented to predict the necessary variation of the hydrogen flow consumption to satisfy the variation of demanded power to the fuel cell. In this research, a hybrid intelligent model was created and validated over a dataset from a fuel cell energy storage system. Obtained results validate the proposal, achieving better performance than other well-known classical regression methods, allowing us to predict the hydrogen consumption with a Mean Absolute Error (MAE) of 3.73 with the validation dataset.

Keywords: fuel cell; hydrogen energy; intelligent systems; hybrid systems; Artificial Neural Networks; power management

---

1. Introduction

Environmental care is currently not only a trend, but it is also an important issue for society and governments. Moreover, for obvious reasons, there is a clear trend in which it is necessary to ensure care for the environment. In point of fact, no impact is very difficult or impossible. But nevertheless, aspects such as sustainability and the maximum possible reduction in environmental impact are very important [1]. In this sense, in terms of energy needs, renewable energies play a key role in contributing to a reduction in environment impact and emissions [2]. However, the impact of the power plant implementation itself, based on renewable sources, has to be taken into account; there is not usually zero impact [3].

Due to it not being possible to achieve the null impact, even with the alternatives and use of renewable energies, there is a legal obligation to optimize and plan installations with maximum efficiency [4]. Furthermore, the efficiency of the facilities must be measured in accordance with the right ratios and criteria with the aim of ensuring the desired minimum impact [5].

Both the greatest environmental impact and the economic investment take place during the construction of the power generation facility. When the plant is in operation, although depending on the technology, it usually needs far less expense. Even in some cases, it is better that the installation be in operation than stopped. When this happens, energy storage becomes a highly recommended solution [6].

Some of the reasons why the electric sector system is very convoluted are as follows [7]. Among all of them, nowadays, the most complex issue is the matching of consumption with demand. If it is added to the different electric energy technologies of generation, the system could be ungovernable and unpredictable. According to some sources, the destabilization could be increased by the renewable energy plants [8].

Due to the above reasons, the energy management systems are an imperative necessity. These possible methods and tools could handle the energy consumption and generation points. It is for these reasons therefore, that the concept of SmartGrids arises [9], in which, among others, the generation and consumption are measured and monitored. With the aim to take decisions, more important than doing the above is to predict the behavior for matching the demand and the generation. Energy storage is a very helpful tool to achieve this goal [10].

There are geographical areas in which buildings have electricity energy needs, and they are not connected to the electric grid [11]. This kind of case does not usually have an easy solution because to connect a building to the electricity network implies a high cost, which is never going to be amortized. A possible alternative, and maybe a more feasible one, is to implement energy storage systems [12].

As shown, in all cases, the energy storage could be a a feasible solution for the mentioned problems [13]. Many are the technologies for this purpose, some of them relatively old and wasteful, like pumping water for its storage [14]. During recent times, due to the energy store need, there are a lot of proposals to solve this necessity [15]. One of the last reasons is the electric car development and trend [16]. Among all existing technologies, the batteries and fuel cells are the most popular ones in this change process [17].

One of the major difficult tasks of storage systems is the efficiency advancement under a global point of view. Nevertheless, under a practical outlook, commonly this efficiency is quantified in economical aspects. Of course, ahead of achieving this moneymaking objective requires hard progress. Lately, there are many proposals with the aim to achieve energy storage methods, systems, processes, and so on, in varied forms. Some examples of this are the following: In [18], an optimal nonlinear controller based on Model Predictive Control (MPC) for a flywheel energy storage system is proposed, in which the constraints on the system states and actuators are taken into account. Ref. [19] describes a system for storing energy deep underwater in concrete spheres, which also can act as moorings for floating wind turbines. A proposal is made in [20] for a deterministic and an interval unit commitment formulation for the co-optimization of controllable generation and PHES (Pumped Hydro Energy Storage), including a representation of the hydraulic constraints of the PHES. The present work is focused on the fuel cells case and, specifically, on the hydrogen-based ones. These are the most common ones due to two basic reasons: hydrogen is a very abundant gas, and it is easy to achieve through a very simple process based on hydrolysis [21].

Given that, as mentioned before, it is very important to obtain the right prediction of both the generation and the consumption, with the aim to achieve the correct decisions [22]. When it will be necessary for energy selling or purchasing, the accurate forecasting must be decisive to be efficient under an economical point of view [23]. Taking into account this affirmation, it is very important to have an effective prediction when a fuel cell system based on hydrogen is used.

For the behavior prediction, accomplishing the process modeling is mandatory. Some of the possible ways for this purpose come from Multiple Regression Analysis (MRA)-based models [24], which have very common limitations in several instances [25]. This problem is due to the possible nonlinearities in most of the cases [26]. A possible alternative that solves this problem is the modeling based on intelligent techniques, with which it is possible to achieve satisfactory results commonly [27]. However, despite the intelligent systems use, bad performance could be possible, if the nonlinearities are several. If it is the case, then, a hybrid system based on clustering techniques, previous to the regression step, frequently gives satisfactory results [28].

A fuel cell performance modeling is accomplishing on the present research work, taking into account some measured parameters at the real storage system plant. Specifically, with the aim to achieve a very useful application, the model must predict the necessary increasing or decreasing of hydrogen flow for a gradual change of provided output power. Remark that the system behavior has a very nonlinear component, whereby clustering techniques are applied. Then, regression based on intelligent techniques are performed.

The rest of the paper is structured as follows. After the present section, the case study is described. Then, it is given the description of the proposed model approach for solving the problem. After that, the results are detailed and the last section exposes the conclusions and the future works.

2. Case Study

The "University of A Coruña" (UDC) has an experimental system to study renewable sources combined with energy storage system and their possibilities to increase the efficiency of the Power Network. Figure 1 shows the basic scheme of this experimental installation. Our research is focused on the energy storage part [29] that uses a fuel cell as a power transformation system to produce electrical energy from the chemical energy stored as hydrogen. The used storage system is a laboratory-size equipment, a system with research purpose.

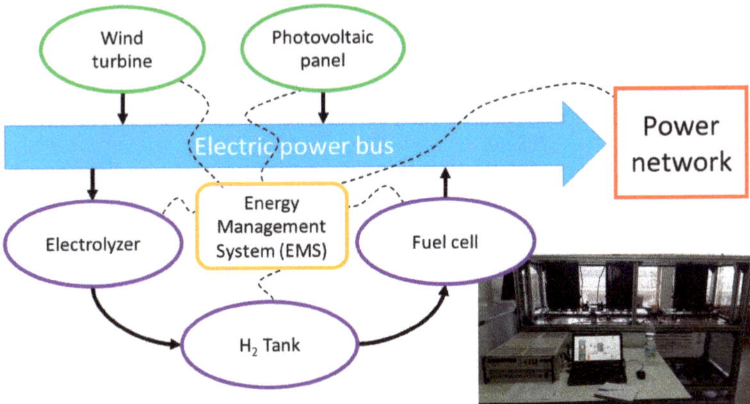

Figure 1. Power system layout.

In Figure 1, the internal diagram of the electric power system is shown. Solar and wind generation and fuel cell are represented as inputs to the power bus; the main output is the Power Network, and also the electrolyzer that demands energy when it produces hydrogen. The Energy Storage System is divided into three different elements: the input Power Transform System (PTS), the electrolyzer; the Central Storage, the H_2 tank; and the output PTS, the fuel cell. The system is controlled and supervised by an Energy Management System (EMS), with the aim of increasing its global efficiency. The Energy Storage System stores energy when the Power Network demand is less than the generation system produces; and the fuel cell produces energy when the Power Network demand is more than the renewable systems could generate.

This research is focused on the output PTS, the fuel cell, and this specific equipment, which are described in detail bellow.

2.1. Fuel Cell

Figure 2 shows the internal scheme of a single fuel cell. The inputs are the H_2 inlet, the fuel input, and the air inlet, the O_2 is picked from the air. The cell has two different outputs: the water outlet, which is the product of the internal reaction; and the H_2 outlet, if the fuel input flow is greater than that of the electrical application needs [30].

The internal reactions can be divided into three different types, depending on the part of the cell: the anode, the cathode, and the global result. Equation (1) shows these three chemical reactions [31,32]. The fuel cell type used in this research is a Proton Exchange Membrane Fuel Cell (PEMFC), one of the most efficient technologies. It has high energy density and low volume and weight against other fuel cells.

$$\left.\begin{array}{ll} \text{Anode:} & H_2 \rightarrow 2H^+ + 2e^- \\ \text{Cathode:} & \frac{1}{2}O_2 + 2H^+ + 2e^- \rightarrow H_2O \end{array}\right\} H_2 + \frac{1}{2}O_2 \rightarrow H_2O + \text{Energy.} \tag{1}$$

Single fuel cells are joined together to create a stack. The stack used is a PEMFC FCgen-1020AVS from Ballard [33], and it is formed by 80 BAM4G polymeric single cells [29]. It has a porous carbon cloth anode and cathode, with a catalyst based on platinum. The whole stack has graphite plates between cells, and aluminum end plates, all of them joined by compression.

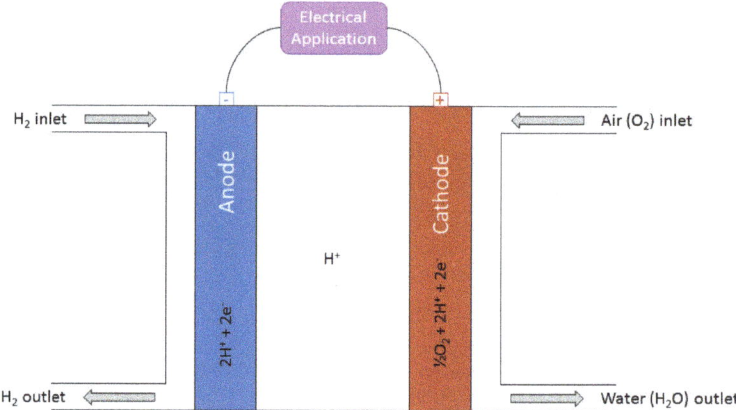

Figure 2. Internal schematic of a fuel cell.

2.2. Power System

The electrical output of the fuel cell is not regulated, the voltage depends on the electrical load and the H_2 inlet flow. However, the electrical applications need a stable input voltage to improve its operation. Figure 3 shows the general power system diagram, where a power converter connects the output of the fuel cell and the electrical application.

It is necessary to emphasize that this kind of power converter controls the output voltage, and the electrical power becomes the most important variable to be controlled. This power depends on the H_2 inlet flow.

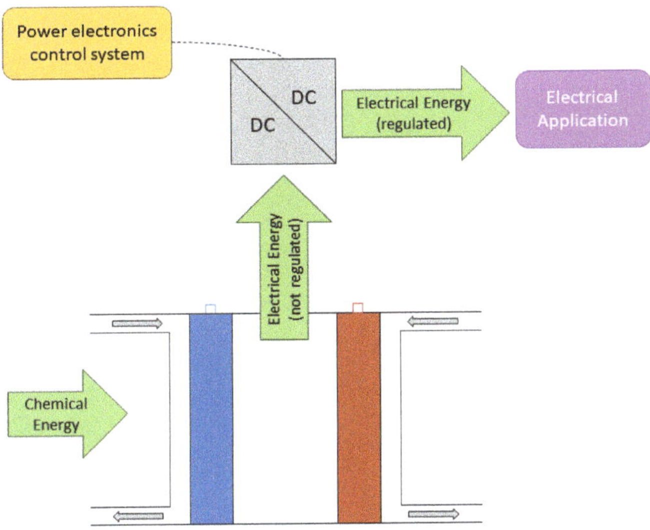

Figure 3. Power system control to stabilize the cell output.

3. Model Approach

Figure 4 shows the basic model of the proposal. Instead of using the hydrogen flow as the model output, the variation in the current flow is predicted. Moreover, as the model is focused in the electrical power produced by the fuel cell, the inputs are the current power, the desired power in the future, and the current H_2 inlet flow. The solution provided in this research lies on the modeling of the necessary fuel flow (hydrogen) for a desired power, to minimized the H_2 outlet of the fuel cell.

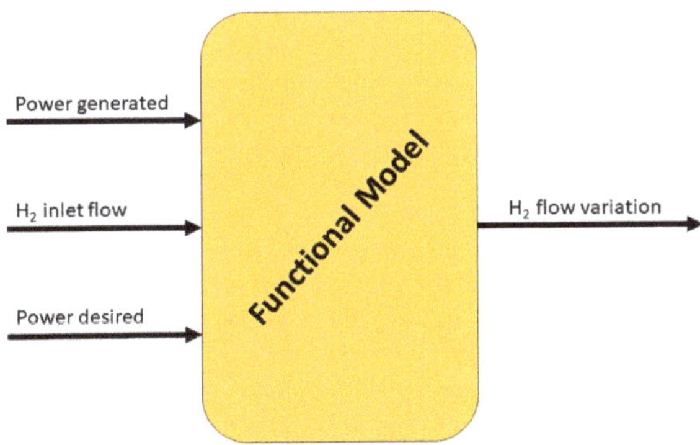

Figure 4. General schema of the functional model.

Figure 5 shows the specific signals and their temporal instants. With the current values of power and hydrogen flow, the model predicts the flow variation two states later to achieve the desired power. As the fuel cell system reacts before the load demands the future power, this model increases the efficiency of the fuel cell.

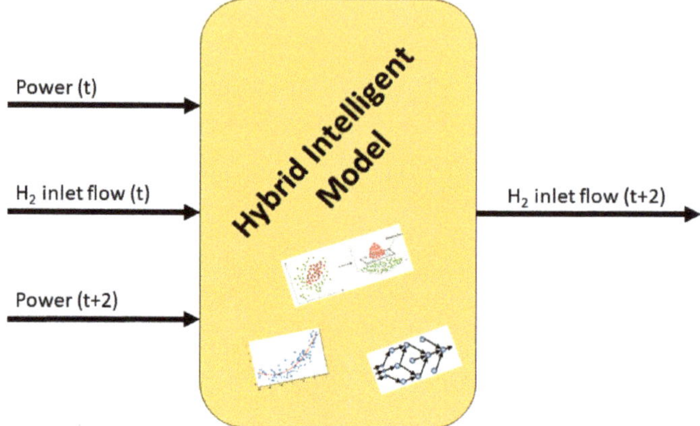

Figure 5. Model approach for forecasting actual current value.

In order to obtain this prediction, a hybrid model has been created using clustering techniques to divide the data into various data subsets. After that, several regression algorithms were trained for each cluster. Figure 6 shows an internal representation of the hybrid model, it can be seen that each group has its own regression model. Each input sample is assigned to a specific cluster, and the output of the whole model will be the output of the specific local model.

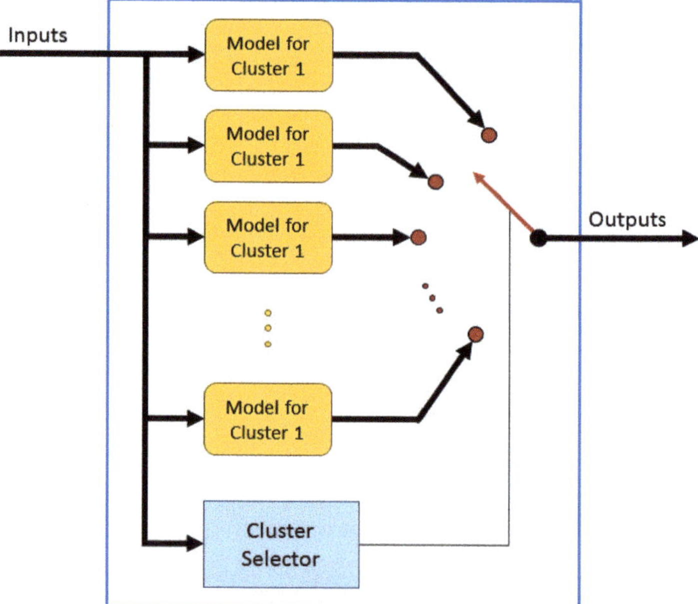

Figure 6. Internal schematic to achieve the hybrid model.

Figure 7 shows the flow diagram followed to create the hybrid model. To perform the third step, the best local model selection, K-Fold cross validation is used to divide the data subsets (cluster data) for training and testing.

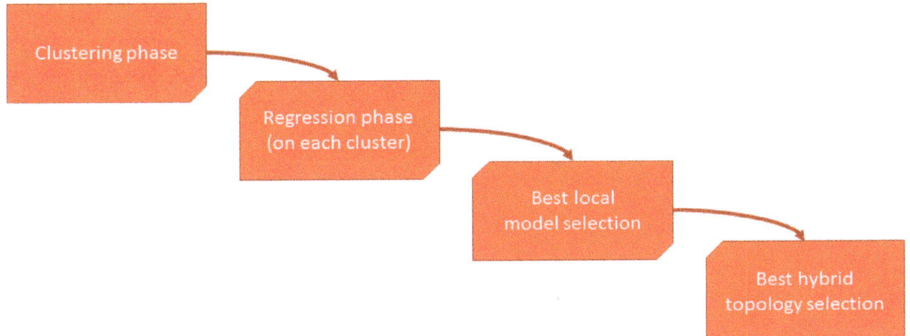

Figure 7. Flowchart of the hybrid model creation phases.

Figure 8 shows this validation procedure. Once K-Fold is selected, *one k-th* of the cluster data is used for testing and the rest for training. With this training data, a regression model is created with the algorithm selected, and the testing data is used to calculate the modeled output. The real testing data output and the predicted one is save in an *Error log*. The training–testing procedure is repeated k times until all the data is used as testing data. At this time, the *Error log* has all the cluster data to calculate the error for each regression algorithm.

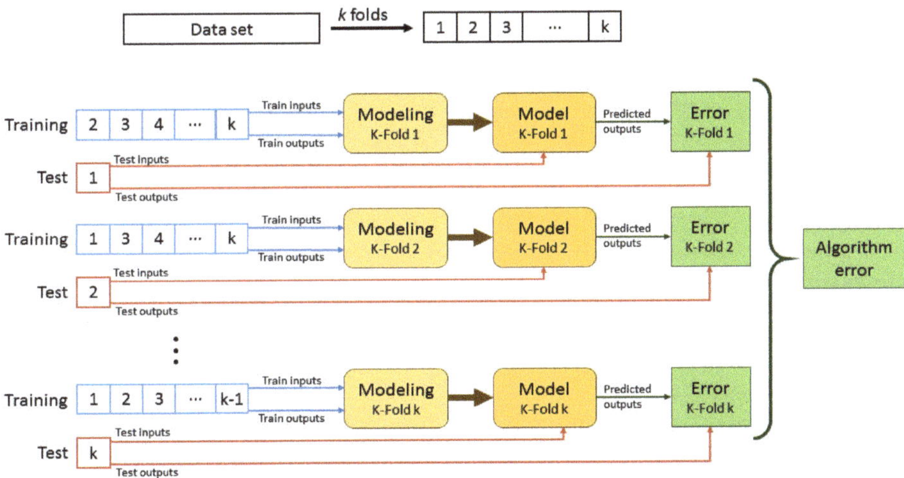

Figure 8. K-Fold training and test data selection.

3.1. Data Processing

To prepare the dataset for the regression phase, a preprocessing of the data is carried out. This process is divided into two different steps. Firstly, the wrong samples are removed—the samples with out of range values. The second step is the normalization, which tries to minimize the training time in the next regression phase. This normalization is based on Max-Min Scaler [34], presented in Equation (2), which obtains new sample values ($Data_j$) in a range from 0 to 1.

$$Data_{j_{new}} = \frac{Data_j - min(Data)}{max(Data) - min(Data)}.$$

(2)

3.2. K-Means Algorithm

The technique chosen for clustering purposes is the K-Means algorithm, in order to define the groups present in a dataset. This algorithm is based on the location of each centroid on its corresponding hyperspace. Thus, the data with a similar nature is situated in the proximity of each centroid, comprising a cluster [35]. The K-Means algorithm tries to define "K" number of centroids. Then, every data point is located in the nearest cluster, always trying to keep the shorter distance (usually the Euclidean) between centroids and each sample.

At the beginning, K-Means implements a training process in order to get the clusters and distribute the data samples. The velocity of this step depends of the numbers of clusters and the size of the dataset. However, the second phase, when each sample data is assigned to its clusters, is done quite fast compared with the initial phase [36].

The procedure to train the K-Means algorithm is explained with this sequence:

- A random set of data samples are chosen as the first set of centroids due to, at the beginning, the center of each group is not known;
- A set of data samples will create a cluster if this set of points are the nearest to this cluster centroid;
- Once the clusters are defined, it is necessary to calculate their associated centroid. These new centroids are chosen as the center of each cluster.

The last two steps are repeated until the centroids are the same two consecutive times. It means that the algorithm is converged and the K-Means algorithm will stop, the new samples can be assigned to its clusters by comparing the distance between the different centroids.

3.3. Artificial Neural Networks

An Artificial Neural Network (ANN) is an artificial intelligence technique based on the biological neurons model; the information is managed by unitary component called a neuron. Like in the biological approach, the artificial neuron is linked with other neurons. Thus, an ANN is able to calculate complex functions thanks to external data input and input from others neurons. The input for each neuron has a weight associated and each neuron has inside an activation function that defines the output.

ANN learning model is based on the fact that this kind of architecture is able to learn from experience thanks to the generalization of cases. Complex functions can be obtained through the training process. The ANN develops a characterization of a problem in order to create an answer in accordance with the input of the problem, without having knowledge about the previous situation. Therefore, the ANN can generalize new solutions from previous ones [37].

The excitation level, also named the output of a neuron, is defined by the activation function [38]. This output can change from 0 to -1 or from -1 to 1. A key feature of an ANN is its topology. It defines how the set of neurons is organized. Thus, the topology includes the 'placement' of the neurons and how they are linked. The architecture of the ANN is defined by four features:

- Number of layers;
- Number of neurons per layer;
- Links between neurons;
- Activation functions.

The Multi Layer Perceptron (MLP) is the basic topology of the ANN. The architecture is organizes as follows: input, a set of hidden layers, output. When the information arrives from the same source to a set of neurons, they belong to the same layer. The information can go from the inputs of ANN or from a previous layer to the next ones. In the MLP, the information from neurons in a layer goes to the same destination—the next layer or the output of the MLP.

Usually, the activation function of the output layer is a specific activation function that depends on the application of the ANN, one of the most common activation functions is the 'linear' one.

3.4. Polynomial Regression

A model defined as summation of several linear functions is known as a polynomial regression model. The amount of basis functions used in the polynomial regression is associated to the number of inputs and is in concordance with the polynomial degree used.

When the first degree is defined, the polynomial can be defined as Equation (3). If the degree increases, the model becomes more complex. A second degree polynomial is shown in Equation (4).

$$f(x) = c_0 + c_1 x_1 + c_2 x_2. \tag{3}$$

$$f(x) = c_0 + c_1 x_1 + c_2 x_2 + c_3 x_1 x_2 + c_4 x_1^2 + c_5 x_2^2. \tag{4}$$

3.5. Support Vector Machines for Regression

The supervised machine learning algorithm known as the Support Vector Machine (SVM) is commonly used for classification. The original SVM algorithm needs only few changes to allow use for regression problems, and this new technique is called Support Vector Regression (SVR). The SVR performs a nonlinear transformation of the original data into a high-dimensional space, and it uses linear regression on this mapping data to calculate the desired output.

In this research, the Least Square SVR (LS-SVR) is used [39], it is a modified algorithm based on SVR that uses the Least Square to minimize the objective function [39]. This modification provides a comparable generalization performance to the SVR [40].

The LS-SVR regression algorithm replaces the insensitive by a classical squared loss function. Equation (5) is used to solve the linear LarushKuhn–Tucker.

$$\begin{bmatrix} 0 & I_n^T \\ I_n & K + \gamma_{-1} I \end{bmatrix} \begin{bmatrix} b_0 \\ b \end{bmatrix} = \begin{bmatrix} 0 \\ y \end{bmatrix}, \tag{5}$$

where:
I_n is a vector of n ones;
T means transpose of a matrix or vector;
γ a weight vector;
b regression vector;
b_0 is the model offset.

LS-SVR only needs to adjust two parameters: the weight vector (γ) and the kernel width (σ) [39].

4. Results

The results of this research are divided into three different parts: the clustering, the regression modeling, and the validation.

4.1. Clustering Results

The number of clusters must be set to train the K-Means clustering technique. As the optimal number of groups is not known previously, in this research, different divisions were trained, from 2 to 9. Table 1 shows the number of samples inside each cluster. Only a maximum of 4 clusters are shown because when K-Means tries to divide the data into more groups, the clusters had few number of samples. Only the clusters with more than 15 samples were saved; the ones with less samples are rejected.

Table 1. Number of samples in each created cluster.

	Cl-1	Cl-2	Cl-3	Cl-4
Global	187			
Hybrid 2	92	95		
Hybrid 3	52	67	68	
Hybrid 4	42	45	47	53

To achieve the best division, the K-Means algorithm was trained 20 times for each number of clusters configuration, each one with random initial centroids. It ensures that the training calculates the best divisions.

4.2. Modeling Results

For the modeling phase, the three different explained algorithms were used for each local model created in the clustering phase.

4.2.1. Artifical Neural Networks

Fifteen different ANN were tested for each group shown in Table 1, all the ANN was trained with the same internal configuration: one single hidden layer, linear activation function in the output layer neuron, and Tan-Sigmoid for the rest of the neurons. The number of neurons in the input layer is fixed to 3, according to the number of inputs; and the output layer has only one, because there is only one output for the model. In the hidden layer, different numbers of neurons were tested, from 1 to 15, and each configuration represents a different model to test in each cluster.

As fifteen configurations were tested, there are fifteen different tables with the error of each configuration calculated with the data test. This research uses 10 K-Fold cross validations to calculate these errors, then, 10 different ANNs were trained before calculating the error. As an example, Table 2 shows the Mean Absolute Error (MAE) when the ANN regression algorithm was chosen, and it is configured with 13 neurons in the hidden layer.

Table 2. Mean Absolute Error (MAE) using Artificial Neural Network (ANN) with 13 neurons in the hidden layer.

	Cl-1	Cl-2	Cl-3	Cl-4
Global	7.7725			
Hybrid 2	43.6784	77.1169		
Hybrid 3	17.8174	26.0211	41.1911	
Hybrid 4	14.8003	25.2539	10.7302	23.8377

4.2.2. Polynomial Regression

In the case of the Polynomial regression algorithm, two different configurations was trained according to the degree of the polynomial used. First- and second-degree were chosen and tested with 10 K-Fold cross validation. The results for Polynomial regression algorithm is shown in Table 3 for the configuration set to first degree.

Table 3. MAE for first-degree Polynomial regression algorithm.

	Cl-1	Cl-2	Cl-3	Cl-4
Global	4.0405			
Hybrid 2	6.5793	13.0030		
Hybrid 3	29.8819	74.4182	14.6885	
Hybrid 4	52.8102	50.7558	53.4093	124.6589

4.2.3. Support Vector Machines for Regression

For the LS-SVR, a Matlab Toolbox by KULeuven-ESAT-SCD was used. This toolbox allows to autotune the internal parameters necessary in the LS-SVR algorithm. The training of this algorithm is made by using this autotune function, and it only generates one model against the others regression algorithm, which creates some different models with different internal configurations. The MAE errors using LS-SVR is shown in Table 4.

Table 4. MAE for Least Square Support Vector Regression (LS-SVR) regression algorithm.

	Cl-1	Cl-2	Cl-3	Cl-4
Global	4.0405			
Hybrid 2	6.5793	13.0030		
Hybrid 3	29.8819	74.4182	14.6885	
Hybrid 4	52.8102	50.7558	53.4093	124.6589

4.2.4. Best Regression Local Models Selection

Table 5 shows the best algorithms for the specific application in this paper. These best algorithms were chosen base on the Mean Squared Error (MSE) values for each cluster. Although the algorithm are the same, the internal parameters are different and each model are adjust to its own dataset. Table 6 shows the MSE value calculate using K-Fold cross validation to test the models.

Table 5. Configuration for each individual hybrid model.

	Cl-1	Cl-2	Cl-3	Cl-4
Global	LS-SVR			
Hybrid 2	LS-SVR	LS-SVR		
Hybrid 3	LS-SVR	LS-SVR	LS-SVR	
Hybrid 4	LS-SVR	LS-SVR	LS-SVR	LS-SVR

Table 6. Mean Squared Error (MSE) for each individual hybrid model.

	Cl-1	Cl-2	Cl-3	Cl-4
Global	18.3348			
Hybrid 2	11.0268	28.0035		
Hybrid 3	20.3464	37.9476	10.7460	
Hybrid 4	65.2756	15.5356	10.2792	94.5818

4.3. Validation Results

To select the best hybrid configuration, a validation dataset is used. This data was separated, and isolated, from the clustering and the regression training phase. Once the best algorithm per cluster is selected, this validation dataset tests the final four configurations of the whole model (a global model and three hybrid models). The results of this validation test is shown in Table 7, and it shows that the best hybrid model is created with three local models.

Table 7. Mean squared error for each model

	Global	Hybrid Model (Local Models)		
		2	3	4
MSE	24.0758	33.8591	23.8121	398.9072

Different error values are calculated with the final configuration to evaluate its performance. The values of these errors are described in the list bellow.

- Mean Squared Error $- MSE = 23.8121$;
- Normalized Mean Squared Error $- NMSE = 0.4438$;

- Mean Absolute Error $- MAE = 3.7318$;
- Mean Absolute Percentage Error $- MAPE = 110.4193$.

5. Conclusions and Future Works

The model created in this research predicts the variation in the hydrogen flow consumption by a fuel cell in an early future. The model uses the desired generated power at the output of the fuel cell, the current generated power, and current hydrogen inlet flow as inputs, and it predicts the variation in the inlet flow as output.

A power converter is used to stabilize the electrical voltage in the output of the fuel cell. It produces the desired voltage for the specific application connected to the fuel cell. As the voltage of the fuel cell varies with the different working points, this power converter allows to control only the electrical power produced by the fuel cell; the output voltage of the converter will be constant all the time.

The bioinspired hybrid model created combines different regression algorithms with clustering to increase the prediction performance of the model. The final model includes three local models with an LS-SVR in each one, and the error values with a validation dataset show that it achieved good results. The NMSE was 0.45, and the MAE was 3.73.

As future works, it is possible to mention the integration of this model as a part of the control system. This configuration would allow to create a kind of predictive control that could increase the efficiency of the fuel cell system, as it would predict the reaction of the system.

Author Contributions: Data curation, J.A.B.-A. and P.N.; Investigation, H.L.G. and J.L.C.-R.; Methodology, H.A.-M. and H.Q.; Project administration, J.L.C.R. and P.N.; Software, H.A.-M. and J.-L.C.-R.; Supervision, E.J. and J.L.C.-R.; Validation, H.Q. and J.A.B.-A.; Writing, original draft, E.J. and H.L.G.

Funding: This work has been funded by Consejería de Educación (Junta de Castilla y León) through the LE078G18 project (UXXI2018/000149. U-220).

Conflicts of Interest: The authors declare no conflict of interest.

References

1. Kuwae, T.; Hori, M. Global Environmental Issues. In *Blue Carbon in Shallow Coastal Ecosystems: Carbon Dynamics, Policy, and Implementation*; Springer: Berlin, Germany, 2019.
2. Karunathilake, H.; Hewage, K.; Mérida, W.; Sadiq, R. Renewable energy selection for net-zero energy communities: Life cycle based decision making under uncertainty. *Renew. Energy* **2019**, *130*, 558–573. [CrossRef]
3. Burduk, A.; Bozejko, W.; Pempera, J.; Musial, K. On the simulated annealing adaptation for tasks transportation optimization. *Logic J. IGPL* **2018**, *26*, 581–592. [CrossRef]
4. Wei, M.; Patadia, S.; Kammen, D.M. Putting renewables and energy efficiency to work: How many jobs can the clean energy industry generate in the US? *Energy Policy* **2010**, *38*, 919–931. [CrossRef]
5. Giacone, E.; Mancò, S. Energy efficiency measurement in industrial processes. *Energy* **2012**, *38*, 331–345. [CrossRef]
6. Dunn, B.; Kamath, H.; Tarascon, J.M. Electrical energy storage for the grid: A battery of choices. *Science* **2011**, *334*, 928–935. [CrossRef]
7. Montero-Sousa, J.A.; Casteleiro-Roca, J.L.; Calvo-Rolle, J.L. Evolution of the electricity sector after the 2nd world war. *DYNA* **2017**, *92*, 280–284.
8. Montero-Sousa, J.A.; Casteleiro-Roca, J.L.; Calvo-Rolle, J.L. The electricity sector since its inception until the second world war. *DYNA* **2017**, *92*, 43–47.
9. de Souza Dutra, M.D.; Anjos, M.F.; Digabel, S.L. A general framework for customized transition to smart homes. *Energy* **2019**, *189*, 116138. [CrossRef]
10. Nizami, M.; Haque, A.; Nguyen, P.; Hossain, M. On the application of Home Energy Management Systems for power grid support. *Energy* **2019**, *188*, 116104. [CrossRef]

11. Jove, E.; Casteleiro-Roca, J.L.; Quintián, H.; Méndez-Pérez, J.A.; Calvo-Rolle, J.L. Anomaly detection based on intelligent techniques over a bicomponent production plant used on wind generator blades manufacturing. *Rev. Iberoam. Autom. Inform. Ind.* **2019**, [CrossRef]

12. Fernandez-Serantes, L.A.; Montero-Sousa, J.A.; Casteleiro-Roca, J.L.; Vilar-Martinez, X.M.; Calvo-Rolle, J.L. Gestión de almacenamiento energético para instalaciones de generación-distribución. *DYNA Ing. Ind.* **2017**, *92*, 140–141. [CrossRef]

13. Good, N.; Ceseña, E.A.M.; Heltorp, C.; Mancarella, P. A transactive energy modelling and assessment framework for demand response business cases in smart distributed multi-energy systems. *Energy* **2019**, *184*, 165–179. [CrossRef]

14. Yang, C.J.; Jackson, R.B. Opportunities and barriers to pumped-hydro energy storage in the United States. *Renew. Sustain. Energy Rev.* **2011**, *15*, 839–844. [CrossRef]

15. Hache, E.; Palle, A. Renewable energy source integration into power networks, research trends and policy implications: A bibliometric and research actors survey analysis. *Energy Policy* **2019**, *124*, 23–35. [CrossRef]

16. Westbrook, M.H. *The Electric Car: Development and Future of Battery, Hybrid and Fuel-Cell Cars*; IET Digital Library: London, UK, 2001.

17. Hall, P.J.; Bain, E.J. Energy-storage technologies and electricity generation. *Energy Policy* **2008**, *36*, 4352–4355. [CrossRef]

18. Ghanaatian, M.; Lotfifard, S. Control of Flywheel Energy Storage Systems in the Presence of Uncertainties. *IEEE Trans. Sustain. Energy* **2019**, *10*, 36–45. [CrossRef]

19. Slocum, A.H.; Fennell, G.E.; Dundar, G.; Hodder, B.G.; Meredith, J.D.C.; Sager, M.A. Ocean Renewable Energy Storage (ORES) System: Analysis of an Undersea Energy Storage Concept. *Proc. IEEE* **2013**, *101*, 906–924. [CrossRef]

20. Bruninx, K.; Dvorkin, Y.; Delarue, E.; Pandžić, H.; D'haeseleer, W.; Kirschen, D.S. Coupling Pumped Hydro Energy Storage With Unit Commitment. *IEEE Trans. Sustain. Energy* **2016**, *7*, 786–796. [CrossRef]

21. Molina-Cabello, M.A.; López-Rubio, E.; M Luque-Baena, R.; Domínguez, E.; Palomo, E.J. Foreground object detection for video surveillance by fuzzy logic based estimation of pixel illumination states. *Logic J. IGPL* **2018**, *26*, 593–604. [CrossRef]

22. Potter, C.W.; Archambault, A.; Westrick, K. Building a smarter smart grid through better renewable energy information. In Proceedings of the 2009 IEEE/PES Power Systems Conference and Exposition, Seattle, WA, USA, 15–18 March 2009; pp. 1–5.

23. Montero-Sousa, J.A.; Fernandez-Serantes, L.A.; Casteleiro-Roca, J.L.; Vilar-Martnez, X.M.; Calvo-Rolle, J.L. Energy storage management for generation-distribution facilities. *DYNA* **2017**, *92*, 140–141.

24. Segovia, F.; Górriz, J.M.; Ramírez, J.; Martinez-Murcia, F.J.; García-Pérez, M. Using deep neural networks along with dimensionality reduction techniques to assist the diagnosis of neurodegenerative disorders. *Logic J. IGPL* **2018**, *26*, 618–628. [CrossRef] [PubMed]

25. Chalki, A.; Koutras, C.D.; Zikos, Y. A quick guided tour to the modal logic S4.2. *Logic J. IGPL* **2018**, *26*, 429–451. [CrossRef]

26. Jove, E.; López, J.A.V.; Fernández-Ibáñez, I.; Casteleiro-Roca, J.L.; Calvo-Rolle, J.L. Hybrid intelligent system topredict the individual academic performance of engineering students. *Int. J. Eng. Educ.* **2018**, *34*, 895–904.

27. Rincon, J.A.; Julian, V.; Carrascosa, C.; Costa, A.; Novais, P. Detecting emotions through non-invasive wearables. *Logic J. IGPL* **2018**, *26*, 605–617. [CrossRef]

28. Casteleiro-Roca, J.L.; Calvo-Rolle, J.L.; Méndez Pérez, J.A.; Roqueñí Gutiérrez, N.; de Cos Juez, F.J. Hybrid intelligent system to perform fault detection on BIS sensor during surgeries. *Sensors* **2017**, *17*, 179. [CrossRef] [PubMed]

29. Mehta, V.; Cooper, J. Review and analysis of PEM fuel cell design and manufacturing. *J. Power Sources* **2003**, *114*, 32–53. [CrossRef]

30. Segura, F.; Bartolucci, V.; Andújar, J. Hardware/software data acquisition system for real time cell temperature monitoring in air-cooled polymer electrolyte fuel cells. *Sensors* **2017**, *17*. [CrossRef]

31. Casteleiro-Roca, J.L.; Barragán, A.J.; Segura, F.; Calvo-Rolle, J.L.; Andújar, J.M. Fuel cell output current prediction with a hybrid intelligent system. *Complexity* **2019**, *2019*. [CrossRef]

32. Casteleiro-Roca, J.L.; Barragán, A.J.; Segura, F.; Calvo-Rolle, J.L.; Andújar, J.M. Intelligent hybrid system for the prediction of the voltage-current characteristic curve of a hydrogen-based fuel cell. *Rev. Iberoam. Autom. Inform. Ind.* **2019**, *16*, 492–501. [CrossRef]

33. Ballard. *FCgen1020-ACS Fuel Cell from Ballard Power Systems*. Available online: https://www.ballard.com/docs/default-source/backup-power-documents/fcgen-1020acs.pdf (accessed on 28 October 2018).
34. Scikit-learn. *Min Max Scaler*; INRIA: Rocquencourt, France, 2018.
35. Orallo, J.; Quintana, M.; Ramírez, C. *Introducción a la Minería de Datos*; Editorial Alhambra S.A.: Madrid, Spain, 2004.
36. Viñuela, P.; León, I. *Redes de Neuronas Artificiales: Un Enfoque Práctico*; Pearson Educación–Prentice Hall: Upper Saddle River, NJ, USA, 2004.
37. Harston, A.M.C.; Pap, R. *Handbook of Neural Computing Applications*; Elsevier Science: Amsterdam, The Netherlands, 2014.
38. del Brío, B.; Molina, A. *Redes Neuronales y Sistemas Borrosos*; Ra-Ma: Madrid, Spain, 2006.
39. Wang, L.; Wu, J. Neural network ensemble model using PPR and LS-SVR for stock et eorecasting. In *International Conference on Intelligent Computing*; Springer: Berlin/Heidelberg, Germany, 2012; pp. 1–8. [CrossRef]
40. Steinwart, I.; Christmann, A. *Support Vector Machines*; Springer Publishing Company: New York, NY, USA, 2008.

Article

Ear Detection and Localization with Convolutional Neural Networks in Natural Images and Videos

William Raveane, Pedro Luis Galdámez and María Angélica González Arrieta *

Departamento de Informática y Automática, Universidad de Salamanca, Plaza de la Merced S/N,
37008 Salamanca, Spain
*   Correspondence: angelica@usal.es

Received: 18 June 2019; Accepted: 11 July 2019; Published: 17 July 2019

Abstract: The difficulty in precisely detecting and locating an ear within an image is the first step to tackle in an ear-based biometric recognition system, a challenge which increases in difficulty when working with variable photographic conditions. This is in part due to the irregular shapes of human ears, but also because of variable lighting conditions and the ever changing profile shape of an ear's projection when photographed. An ear detection system involving multiple convolutional neural networks and a detection grouping algorithm is proposed to identify the presence and location of an ear in a given input image. The proposed method matches the performance of other methods when analyzed against clean and purpose-shot photographs, reaching an accuracy of upwards of 98%, but clearly outperforms them with a rate of over 86% when the system is subjected to non-cooperative natural images where the subject appears in challenging orientations and photographic conditions.

Keywords: ear detection; computer vision; convolutional neural network; image recognition; video analysis

---

1. Introduction

The problem of people recognition by means of identifying them biometrically by their ear has received considerable attention in the literature. Forensic science has often used a person's ear to establish someone's identity, and considerable improvements are being made in this field to improve these systems—more so now that it starts to be implemented as a new method for biometric recognition [1]. However, for an ear recognition system to be accurate, the first and obvious step it must take is to properly detect the presence and location of an ear within an image frame. This seemingly simple task is often made more difficult because in practice, such images very commonly present the subject's ear in poses which are much different to those a system is usually trained for. Furthermore, occlusion and partially visible ears is very common in natural images, and it presents a challenge which must be addressed.

The Convolutional Neural Network (CNN) [2] is considered today to be one of the broadest and most adaptable visual recognition systems, especially in the case where the imagery is highly variable in form, illumination, and even perspective. A standard CNN is made up two sequential parts, the first one is in charge of feature extraction and learning based on these features, while the second one is (usually) dedicated to classification and the final recognition of the object of interest. A gradient descent algorithm [3] can be used to train these two stages together, end-to-end, and it is precisely this characteristic which gives CNNs their power and flexiblity. This type of networks have, in recent years, come to almost entirely replace other machine learning systems. This is especially the case in image recognition tasks over large datasets [4]. These systems are even capable of performing better than humans can when manually classifying large image datasets [5]. In this work, we exploit the flexible architecture of CNNs to apply them in a custom-designed manner to the particular task of human ear recognition.

The article follows this outline: Section 2 presents a review on the existing methods for the detection of ears and describes the current state of the art. A brief review and explanation of typical CNN architectures is also given. Section 3 describes the methodology our proposed system follows; Section 4 discusses the results and compares them qualitatively to existing methods; and finally Section 5 gives our conclusions and discusses future lines of work that will follow from this research.

2. Background

2.1. Ear Detection State of the Art

Most systems that do ear detection rely on properties in the geometry and morphology of the ear, such as in specific features being visible, or patterns in frequency of low level features. Considerable progress has been made recently in the area of biometrics related to the human ear. One of the best known techniques for ear detection was given by Burge and Burger [6] who proposed a system that makes use of deforming contours, although it does need user input for initializing a contour. As a result, the localization process with this system is not truly automated. Hurley et al. [7] uses force fields, and in this process the location of the ear is not necessary as input in order to do the recognition; however, this technique is very sensitive to noise and requires a clean image of the ear to perform well. In [8], Yan and Bowyer uses a technique that requires two user defined lines to carry out the detection, which again is not fully automated—as one of the input lines must run along the boudnary between the ear and the face, and the second line must cross vertically through the ear, thereby providing a rough localization of the ear as input to the system.

Three additional techniques are given by Chen and Bhanu for the task of ear detection. First of all, they develop a classifying system that can recognize a varying shape indices [9]. This technique, however, only works on images of a side view of the face and is furthermore not very robust against variations in perspective or scale. They also proposed a system that analyzes individual image patches that exhibit a large amount of local curvature. This system makes use of "Step Edge Magnitude", as the technique is called [10]. This system is template-based, requiring a stencil for the usual outline shape of the helix and anti-helix of the ear, this template is then fitted to line clusters. One final technique they propsed reduces the possible number of ear detection candidates by detecting patches of skin texture as an initial step before applying a similar helix stencil matching system to the local curvatures [11].

Another example for detection is described by Attrachi et al. [12] who use contour lines to detect the ear. They locate the outer contour by performing a search on the image for the longest single connected edge feature in the image. By selecting three keypoints for the top, bottom, and left of the localized region. Image alignment can then be done by forming a triangle, such that its barycenter can be used as alignment reference. A. Cummings et al. [13] propse a techinque based on image ray transform that finds the specific tubular shape of an ear. This system relies on the helical/elliptical shape of the ear for localizing it. Kumar et al. [14] created a technique that starts by segmenting the skin, then creates an edge map with which it can finally localize the ear within the input image. They then proceed to use active contours [15] to get a more precise location of each contour.

While there are many proposals attempting to solve the problem of ear detection, only a small portion of them has been described here. An overview is presented in Table 1 outlining the best known methods, along with their reported accuracy rates, when available. A deeper review is also given in [16].

Table 1. Existing ear detection approaches.

Publication	Detection Approach	Database Size	Accuracy Rate (%)
Abaza et al. [17]	Cascaded Adaboost	940	88.72
Ansari and Gupta [18]	Edge Detection and Curvature Estimation	700	93.34
Alvarez et al. [19]	Ovoid Model	*N/A*	*N/A*
Arbab-Zavar & Nixon [20]	Hough Transform	942	91
Arbab-Zavar & Nixon [21]	Log-Gabor Filters and Wavelet Transform	252	88.4
Attarchi et al. [12]	Edge Detection and Line Tracing	308	98.05
Chen & Bhanu [9]	Template Matching with Shape Index Histograms	60	91.5
Cummings et al. [13]	Ray Transform	252	98.4
Islam et al. [22]	Adaboost	942	99.89
Jeges & Mate [23]	Edge Orientation Pattern	330	100
Kumar et al. [14]	Edge Clustering and Active Contours	700	94.29
Liu & Liu [24]	Adaboost and Skin Color Filtering	50	96
Prakash & Gupta [25]	Skin Color and Graph Matching	1780	96.63
Shih et al. [26]	Arc-Masking and AdaBoost	376	100
Yan & Bowyer [27]	Concha Detection and ActiveContours	415	97.6
Yuan & Mu [28]	CAMSHIFT and a Contour Fitting	*Video*	*N/A*

An issue to consider is the great importance of robustness against pose variation and occlusion when an ear detection algorithm is put to practice. It is worthwhile to note that most of the detection systems listed above are not tested nor developed for difficult occlusion scenarios, such as partial occlusion by the hair, jewelry, or even hats and other accessories. The most likely reason is simply the lack of public datasets containing appropriately occluded images. Furthermore, to the best of our knowledge, there is no major research that has been performed on the effect of ear occlusion in natural images.

Additionally, there does not seem to exist any approaches for the specific task of ear detection based on CNNs. Not surprisingly, as CNNs have only started to become popular relatively recently, and the extent of biometric applications using this type of system has so far been limited to full face detection, for example [29].

2.2. Convolutional Neural Networks and Shared Maps

This work is based mainly on a neural network that does classification as its main task. This is a standard CNN with an architecture composed of convolutional and max-pooling layers in alternating order as part of the feature extractor stage. After this, a few fully connected linear layers make up the the final classification network stage.

The network's first/input layer always consists of at one or more units that contain the input image data to be analyzed. For this task, the input consists of a single grayscale channel as input data to the system.

Data next travels to each of the feature extraction stages. The first part of every such stage is a convolutional layer, wherein each neuron linearly combines the convolution of one or more maps from the preceding layer, and then passes the output through a nonlinearity function such as $tanh(x)$. A convolutional layer is usually paired with a max-pooling layer which primarily reduces the dimensionality of the data. A neuron in this type of layer acts on a single map from the corresponding incoming convolutional neuron of the previous layer, and its task is to pool several adjacent values in the map for every sampling pixel in the neuron. The sampling function used takes the maximum value among the pooled region.

The information then travels to one or more additional feature extraction stages, each of which works in a very similar manner as that described above. The result of this is that every stage extracts more and more abstract features that can eventually be used to classify the input, a process done in the final stage of the network. This consists of linear layers which ultimately classify the extracted features on the previous layered stages through a linear combination similar to a traditional multi-layer perceptron.

At the end, the output of the final layer doing the classification finally selects the class that best matches the input data image, based on the predetermined annotation labels with which the system was trained. The output of the network is composed of multiple numeric values, each one giving a probability-like expectancy of the image belonging to the particular class associated with each corresponding estimate.

Recognition of images with dimensions bigger than the input data size with which a CNN was trained with can be achieved by using sliding windows. This is defined by two parameters: S is the size of the window to use, which is set to the network's original input data size; T is the window stride, a value that specifies how far apart sequential windows are spaced. As a result, the stride parameter defines the number of individual windows that must be analyzed for a given input. It is therefore necessary to choose an optimal value for the stride, since this amount is inversely proportional to the classifier "resolution", in other words the resolving power of fine featues in the image. The resolution, in turn, also determines the computing resources necessary to analyze the number of windows W, as more windows obviously require more computations. For an image of size $I_w \times I_h$, the number of windows is determined as follows:

$$W = \left(\frac{I_w - S}{T} + 1\right)\left(\frac{I_h - S}{T} + 1\right) \quad \Longrightarrow \quad W \propto \frac{I_w I_h}{T^2} \tag{1}$$

As an example: Taking an input image that has been downsampled to 640×360, individual windows can be defined, each one of size $S = 64$. To simplify calculations, a stride value of $T = S/2 = 32$ can be used. In this case, a network would require 190 executions to fully analyze each extracted window at this scale. If a smaller stride is used, the computation requirement increases. For example reducing the stride to $T = S/8 = 8$, results in over 2700 individual CNN executions. Taking into account that a single CNN execution, due to its complex nature, can require several million floating point operations, it can be seen that a dense window stride value can increase exponentially the computing toll on the system.

This process can be greatly optimized by executing the network as Shared Maps, a detailed explanation of which is given in [30]. This allows executing the network for the entire image frame in parallel, thus requiring a single execution. Although, a shared map execution of the CNN is higher in computational cost than that of a single window, it can still save on the total computing resources required for the full image by not requiring to re-analyze overlapping regions of adjacent windows, resulting in speed-ups of up to 30x. This process is exploited at its fullest potential here, and its implications are taken into account when designing the structure of the network for this task, as will be described later in this work.

3. System Description

3.1. Datasets

The existence of ear-centric data is limited and sparse. There exist no standard datasets upon which a large body of work can be contrasted with. As a result, there is great difficulty in properly comparing the system we propose with those described in Section 2, as they primarily use private data.

In this work, however, we attempt to use a variety of datasets in order to establish some benchmarks upon which future works can be built upon. For this purpose, we use a total of four datasets in our experiments. Three of these are public and only one is private. Each of these datasets has a set of features which make them particularly useful for a particular task, and each one introduces new challenges. As such, we use them all to base a selection of real-world experiments on each.

Table 2 gives an overview of the content in each dataset, and Figure 1 displays some samples of each to qualitatively demonstrate their contents.

Table 2. Details on the contents of the various datasets used in this work.

Dataset	Dataset Size	Subjects	Images per Subject	Resolution Size *pixels*	Color Channels	Content	Source
AMI [31]	700	100	7	492×702	*Color*	Closeup ears, both sides	*Photo*
UND [32,33]	464	114	4	1200×1600	*Color*	Bust profile, right side only	*Photo*
Videos (Train)	950	5	190	1920×1080	*Color*	Head profile, both sides	*Video*
Videos (Test)	910	7	130	1920×1080	*Color*	Head profile, both sides	*Video*
UBEAR v1.0 [34] (Train)	4497	127	35	1280×960	*Grayscale*	Head profile, both sides, and masks	*Video*
UBEAR v1.1 [34] (Test)	4624	115	40	1280×960	*Grayscale*	Head profile, both sides	*Video*

The first dataset is the AMI dataset [31], a collection of 700 closeup images of ears. These are all high quality images of ears perfectly aligned and centered in the image frame, as well as having high photographic quality, in good illumination conditions and all in good focus. This dataset is therefore exemplary in order to test the recognition sensitivity towards different ears, however, due to the closeup nature of the images, they are not really well suited for ear localization tasks.

The second dataset we use is the UND dataset [32,33]. A collection of photographs of multiple subjects in profile, where the ear covers only a small portion of the image. The photographic quality of these images is very high, and again all in constant and good illumination, and with none of the ears being occluded by hair or other objects. The poses of subjects varies very slightly in relation to the camera, but not so much as to introduce distracting effects due to head rotation and pose. As a result, these images are suitable in testing the specific task of localization among a large image frame, while avoiding the challenges of viewpoint and illumination variation.

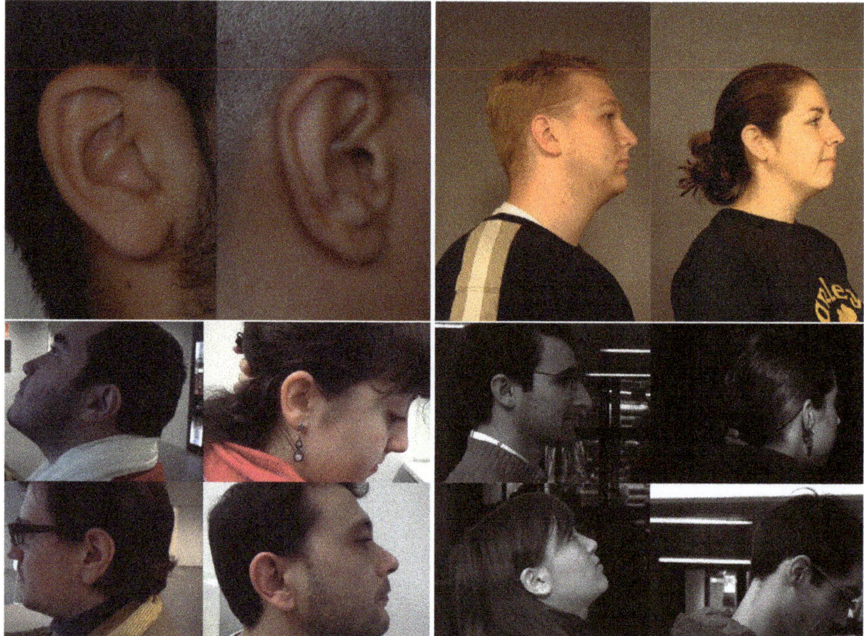

Figure 1. Samples from each of the four datasets used in this work: (**Top Left**) AMI Dataset, (**Top Right**) UND Dataset, (**Bottom Left**) Videos Dataset, (**Bottom Right**) UBEAR Dataset.

The third dataset is the Video dataset. A private collection of 940 images composed of HD frames extracted from short video sequences of voluntary participants. There are 14 image sequences of 7 subjects—one for each person's ear. Each sequence consists of 65 frames from a span of approximately 15 seconds in time extracted from a continuous video. The subjects were asked to rotate their heads in various natural poses following smooth and continuous motions throughout the sequence. The illumination and environment are relatively consistent across all videos, and subjects were asked to move any potential occlusion away from their ears. We use this dataset primarily to test the detector sensitivity only towards different relative rotations of the subject's head in relation to the camera, while avoiding challenges due to variable illumination. The higher number of images per subject, combined with a low number of total subjects, are useful to also reduce the effect from using a large number of wildly variable ear shapes in the tests, and again, concentrate mainly on their pose. A variation of this dataset was created and set aside for training purposes. This comprised profile image frames from an additional 5 participants, different from the subjects in the test dataset.

The final and perhaps most important dataset we use is the UBEAR [34] dataset. This is a very large collection of images of subjects shot under a wide array of variations, which spans multiple dimensions—not only in pose and rotation, but also in illumination, occlusion, and even camera focus. These images, therefore, simulate to a very good degree the conditions of photographs in non-cooperative environments were natural images of people would be captured ad hoc and used to carry out such a detection. These images, although definitely being ear-centric, make no attempt at framing or capturing the ear under perfect conditions, and as such reflect a real-world test scenario. As our main interest in this work is the detection of ears in natural images, this then becomes our main dataset to test the fullest potential of the system we propose. Table 3 gives a more in depth review on the different challenges found in this specific dataset.

Table 3. Differences and challenges presented in the UBEAR dataset.

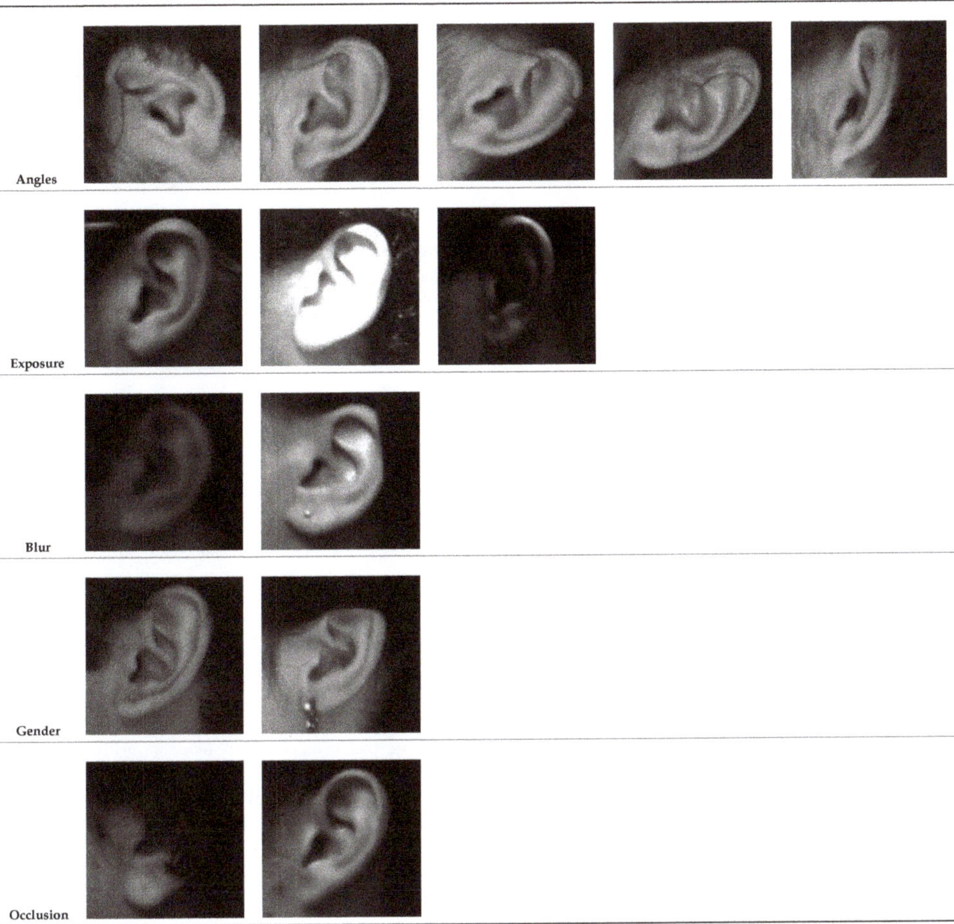

It is also important to note that the UBEAR dataset comes in two versions, both of which consist of unique non-repeating images across both sets. The first of these versions, named 1.0, includes a ground truth mask outlining the exact location of the ear in each image. As will be described later, this inclusion was important for our training procedure. The other version, 1.1, does not include such masks, and is therefore reserved for testing and experimentation.

3.2. Convolutional Neural Network

The CNN used is based on a standard architecture with a few customizations made to the architecture which greatly help for the use case presented. The network architecture used is visually depicted in Figure 2.

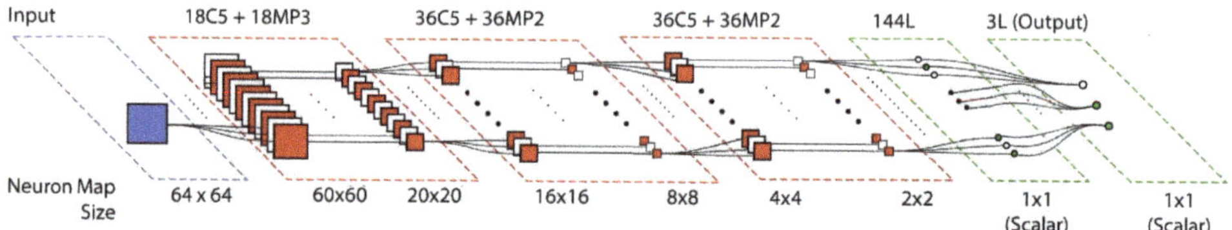

Figure 2. Convolutional Neural Network (CNN) Architecture used in the system.

The target use case of the system is to perform real time ear detection, especially with input video streams. For this, a system that can run quickly is a fundamental requirement. For this reason, an optimized architecture is needed. The target classes we seek to recognize with the neural network are only three: (i) Left Ear, (ii) Right Ear, and (iii) Background—referred to by their corresponding abbreviations: LE, RE, and BG in all the following descriptions of the system. As the data variability within each class is relatively low, with many training data samples having a similar set of characteristic ear features, the network can perform relatively well by learning only a small number of unique features (unlike the case of large modern CNNs). Therefore, a small neural network, with a low layer and neuron count is enough to learn the training data used by this system.

Furthermore, a size of 64×64 is selected for the input data of the network, as images at this size carry enough features and information to properly define the ear shape, while at the same time not being so large that the system would require large convolutional kernels to properly analyze the images.

Finally, as Shared Maps execution will be used to do the analysis over full images, the maximum accumulated pooling factor needs to be kept small. This ensures that the stride size on the final output map is still small for fine localization to take place. For this reason, 3 convolutional and pooling layers are decided as the base of the architecture.

Knowing these three constraints, for the input and output, and the maximum number of layers, through a process of iterative trial and error, a final architecture was decided upon as follows:

$$18C5:MP3 + 36C5:MP2 + 36C5:MP2 + 144L + 3L$$

where the notation $A(C, MP, L)B$ means a convolutional (C), max-pooling (MP) or linear (L) layer, of A neurons, and kernel size B. This architecture, when executed as Shared Maps, yields a minimum window stride size of $3 \times 2 \times 2 = 12$, which is quite efficient for purposes of detection over a half-HD image frame, as it allows analyzing the image at intervals as close as 12 pixels apart, or multiples thereof.

3.3. 3-CNN Inference

Training a single neural network and expecting it to be sufficient to properly tell apart ears from background noise in real-world imagery is quite the leap of faith.

In practice, a neural network of this type will be quite capable at properly recognizing the large majority of ear-shaped objects that are presented to it. Thus, when tested against a set of cut-out ear images specifically prepared for such a task of recognition, its true positive inference performance will be quite good. However, it will be prone to make many mistakes when presented with background images or noise. The network is trained with a BG class to help it learn the difference between an ear and background noise, but no matter how the training for this class is prepared, a CNN will always be prone to false detections simply due to the internal functionality of neural networks. There will always be patterns or combination of features that can be easily found on natural imagery which will randomly trigger internal neural paths and thus produce a large false positive rate as well—a type of artificial pareidolia. For real world purposes of image detection over large input image frames, this results in a large number of false hits. Table 4 describes this effect in more detail. A single CNN will very often detect the ear correctly (Ears Detected metric), both in close up images as in the AMI dataset (99.70%), but also in the more challenging full image frames of the UBEAR dataset (93.90%). However, this metric disregards the effect of false positives. The F1 metric is useful to uncover the great performance disparity that occurs in reality. While, in the AMI dataset, the F1 value remains high (99.86%), in the UBEAR dataset it drops abysmally (41.46%) due to the very large number of false positives introduced.

Table 4. Single vs. 3-Convolutional Neural Network (CNN) inference performance, showing how both systems vary greatly when tested against different data types.

Dataset	Algorithm	True Positive	False Positive	False Negative	Ears Detected (%)	F1 Score (%)
AMI	Single CNN	698	0	2	99.70	99.86
	3-CNN	693	0	7	99.00	99.50
UBEAR	Single CNN	4326	11,935	280	93.90	41.46
	3-CNN	3814	605	661	82.80	85.77

This problem can usually be addressed by creating ensemble systems consisting of multiple classifiers, each one different in a specific manner. They all analyze the same data input, and their different outputs are then combined to create a final result whose accuracy will usually be larger than that of any single classifier running by itself [4].

We apply a variation on this idea, in that we do not process all classifiers in the ensemble with the exact same input data, but rather we present different data to each component of the ensemble. Therefore, each of the classifiers must then be trained to specialize in the kind of data which will be presented to it. The different data inputs are carefully constructed so that each one carries meaning specific to that component according to its own specialization.

The main idea then is to feed to three neural networks three different images, each one corresponding to the same image region being analyzed but at different cropping scales. Figure 3 depicts the three different scales which are ingested by the triple classifier ensemble. We appropriately label each of the three networks used to analyze these as S, M, and L (for their corresponding size abbreviations).

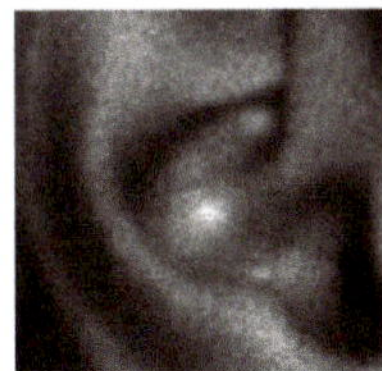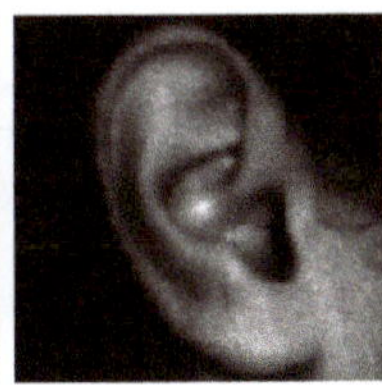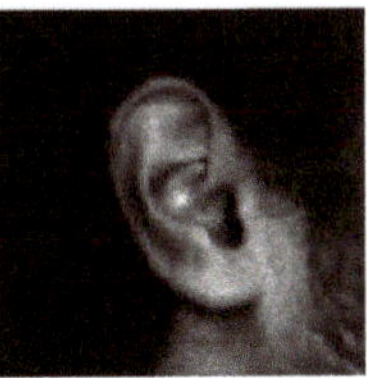

Figure 3. The three scales that are used for every data point in the training dataset.

The purpose of the three scales is mainly to train specialized networks for the specific purposes of (i) recognizing the tubular features of the inner ear, (ii) framing the correct coordinates of the ear, and (iii) inferring ear context within a surrounding head region. Training a network with any single one of these scales would specialize it in that particular data, but the network would be oblivious to other natural image data with similar structure but not really belonging to a true ear, and thus leading it to produce a large number of false positives which would end up affecting the overall detection accuracy. However, the three networks working together as a committee of classifiers produces a much more robust result that is far more resilient against noise, as a true positive hit will require the activation of all three networks, simply by integrating contextual information into the system.

Each of the three neural networks produces three output values, which correspond to the likelihood of each target class having been perceived in that network's input. We denote the output values as O_A^K, where $A \in \{S, M, L\}$ represents the network index denoted by its size, and $K \in \{LE, RE, BG\}$ represents the output class index of each network, for each of the possible detection outcomes. Each of these outputs will lie in the $[-1, +1]$ range as the neural networks have been trained with those ideal values.

To combine the outputs of all three networks as a unified ensemble, we filter each class output with the corresponding values across all three networks, after each one has been linearly rectified. The final outputs of the ensemble are defined by:

$$O_F^{LE} = \lceil O_S^{LE} \rceil^+ \cdot \lceil O_M^{LE} \rceil^+ \cdot \lceil O_L^{LE} \rceil^+ \tag{2}$$

$$O_F^{RE} = \lceil O_S^{RE} \rceil^+ \cdot \lceil O_M^{RE} \rceil^+ \cdot \lceil O_L^{RE} \rceil^+ \tag{3}$$

$$O_F^{BG} = \lceil O_S^{BG} \rceil^+ \cdot \lceil O_M^{BG} \rceil^+ \cdot \lceil O_L^{BG} \rceil^+ \tag{4}$$

where $\lceil x \rceil^+ \equiv max(0, x)$, is a linear rectification operation. By passing through only the positive values of each interim output, we avoid interference from multiple negative values, any of which then has the effect of zeroing the final output. Figure 4 depicts the process visually.

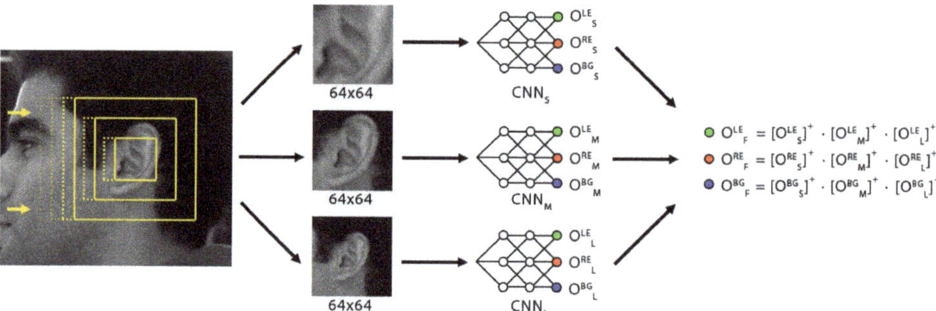

Figure 4. Data flow in the inference process of 3-CNN detections.

The net effect of this process, then, is to have all three networks work in tandem, where only the regions for which all three networks are in full agreement will survive. Furthermore, the final output will be weighed by the individual network certainty, and thus regions where all three networks have a high likelihood output will outweigh regions where the output distribution is more disparate.

3.4. Training Data

As the system will comprise three individual neural networks, we already know beforehand that the training data will need to be gathered in accordance to the requirements for each of the individual networks.

It was previously discussed that each network will essentially analyze three different crop sizes of each region, so the data for all three can be prepared simultaneously by simply starting with one dataset, and extending it by cropping and scaling accordingly to generate the data for the two other sizes.

Existing image datasets consisting of segmented ear photographs are very scarce and small in volume. Creating sufficiently large amounts of training data, therefore, required a lot of manual labor in image manipulation. The datset UBEAR v1.0 was particularly helpful, as described in Section 3.1, in that it includes for each of its images a ground truth mask. This mask outlines the exact location of the ear in each image, and this aided in cropping out the corresponding bounding boxes for each ear. Not all patches from this dataset could be used, however, as many were extremely blurry and not appropriate for training. In the end, approximately 3000 images were used from this set for training data.

Furthermore, we supplemented the training data with additional samples that were manually cropped from video frames. These originate from the Videos (Train) dataset described in Section 3.1. With this addition, the training dataset now consisted of roughly 4000 images.

To increase the dataset size even more, the data was augmented in two ways: (i) images were randomly modified by adding small translations, rotations and rescales; and (ii) images were horizontally flipped, and the resulting image was assigned to the opposite ear dataset. This artificial augmentation boosted the training data size tenfold. It now consisted of approximately 40,000 images, or 20,000 for each ear.

In order to prepare for the training of our final 3-CNN architecture, we processed the images for each ear side into three separate sets, for each of the 3-CNN scales: S, M, and L. This was done by simply cropping and rescaling each sample appropriately.

The process was repeated for both sides, thus producing six separate image collections for left and right ears, at each of the three scales. Finally, one more background noise dataset was also created, of the same size as the others, and consisting of randomly cropped patches from a large flickr photo database and from non-ear regions of the UBEAR and Videos training sets.

In total, we ended up with seven distinct collections for training purposes, each one consisting of roughly 20,000 images. Figure 5 shows an example of these.

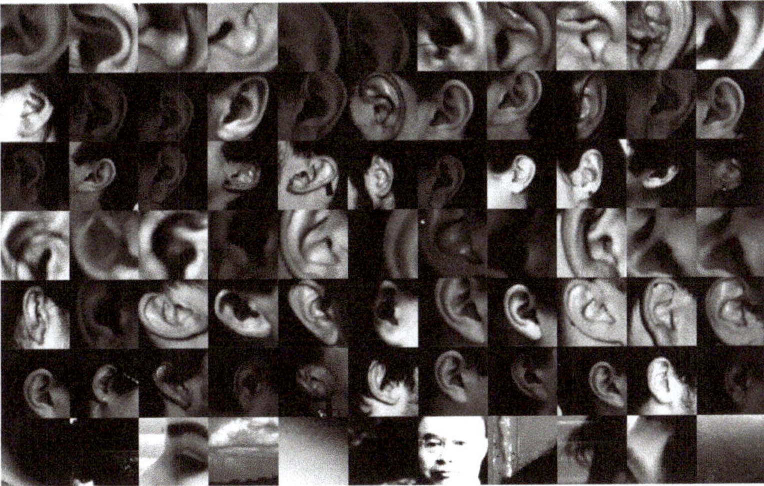

Figure 5. A small subset of each of the seven datasets used for training. From top to bottom: Left-Small, Left-Medium, Left-Large, Right-Small, Right-Medium, Right-Large, Background.

3.5. Network Training

Our final neural network classifier was trained with the three-scale collection described above. Each of the three networks used a 3-class training dataset compiled from left and right ears at the corresponding network scale, and a copy of the background image collection.

The structure of all three networks was exactly the same, and is the one described in Section 3.2. The input consists of a single grayscale channel image resized to a square of size 64×64. The input images are then passed through a pre-processing step which consists of a Spatial Contrastive Normalization (SCN) process, which helps to enhance image edges and redistribute the mean value and data range, something which greatly aids in the training of CNNs.

Each network is trained with its corresponding small, medium or large datasets. A standard SGD approach was used for training, and ran for a duration of approximately 24 iterations until no further improvement could be made on the test-fold of the data. Ideal targets for each of the output labels were assigned in the $[-1, +1]$ range, where active labels are positive, and inactive labels are negative. This distribution was chosen in this manner (as opposed to the more traditional $[0, 1]$ range) to aid with the 3-CNN inference as explained in Section 3.3.

All datasets are divided into training and testing folds, at an 80% to 20% ratio as per standard machine learning training practices. The final results of training over these two sets are summarized in Tables 5 and 6.

Table 5. Final confusion matrix of the training data fold.

Classified As/ Real Class	Left Ear	Right Ear	Background	Total in Class	Accuracy (%)
Left Ear	16,040	56	88	16,184	99.11
Right Ear	46	16,064	74	16,184	99.26
Background	63	194	15,927	16,184	98.41
			Total	48,552	98.93

Table 6. Final confusion matrix of the testing data fold.

Classified As/ Real Class	Left Ear	Right Ear	Background	Total in Class	Accuracy (%)
Left Ear	3964	34	49	4047	97.95
Right Ear	14	4002	31	4047	98.89
Background	8	42	3997	4047	98.77
			Total	12,141	98.54

3.6. Detection

Runtime operation of the network is performed through Shared Map execution of CNNs. This allows for an optimized method of inferring detection predictions from a full image frame in a manner that is much more efficient than the traditional sliding window approach.

The process requires the input image to be first prepared as a multi-scale pyramid. This is simply to be able to detect ears in all possible sizes relative to the image frame, so as to be able to properly carry out the detection, regardless of the subject's relative distance to the camera.

Each of these pyramid levels will be given to each of the three networks to be analyzed independently. Each network, thus, creates three output maps per level, corresponding to each of the target classes trained, LE, RE, and BG. Figure 6 depicts the shared map execution of one of the networks for a particular pyramid level of size 274×366.

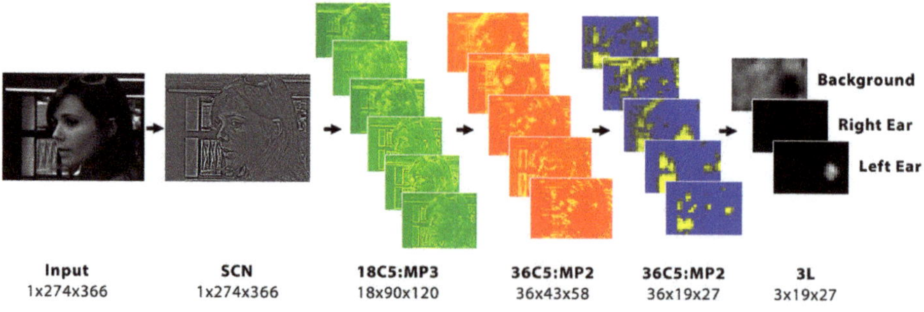

Figure 6. Shared map execution of one of the CNNs over a sample input image.

Every pixel in each of these output maps corresponds to that class' predicted likelihood at a window whose location can be traced back to the input image according to the shared map's alignment and position configuration. Figure 7 shows how windows can be re-constructed from these shared maps and they correspond precisely to the multiple detections that a traditional sliding window approach would produce, but at a fraction of the computing time.

Figure 7. Sample of multiple overlapping detections casted as individual detection windows on an input image.

In order to collapse these multiple detections into a single final result, a partitioning algorithm based on Disjoint-set data structures is used. This is very similar to the *groupRectangles* and *partition* functions of OpenCV [35], but customized in a few particular ways. This algorithm allows the grouping of similarly positioned and scaled windows as all belonging to a single object detection. Figure 8 shows a diagram of how the grouping algorithm would behave on various sample window clusters.

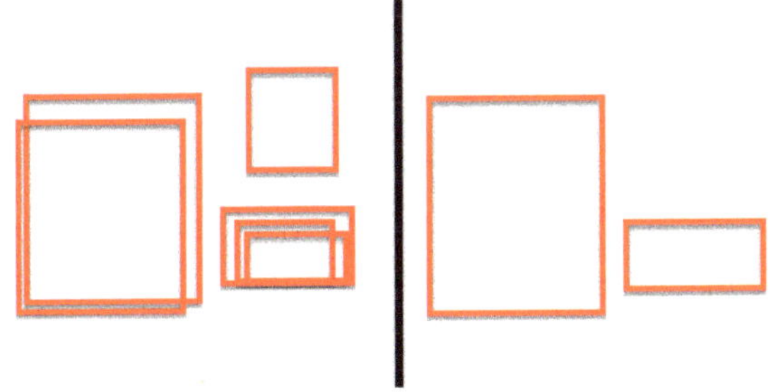

Figure 8. Sample of how the partitioning and grouping algorithms cleans up multiple overlapping detection windows.

This is a very common practice taken as a post-processing cleanup procedure in many computer vision tasks. For this particular work, however, a special grouping rule is created in order to weigh the grouping allowance.

For each of the two positive classes, LE and RE, the following procedure is performed:

Every window i has a value assigned to it corresponding to the neural network output prediction value at that window, denoted by O_i. This window weighs its own value by squaring itself. Therefore, windows with a low prediction value have their overall importance reduced, whereas windows with a large output value to begin with, maintain their standing in the grouping.

For a potential grouping cluster j composed of N multiple windows, each with a weighed output value O_i^2, the final output value G_j for the group is then given by:

$$G_j = \sqrt{\frac{\Sigma_i^N O_i^2}{N}} \tag{5}$$

This corresponds to an RMS of all composing window output values in that cluster. The end result of this is that the process favors those clusters that are composed of windows with large significant confidence outputs, where as windows with low confidence (such as in the case of false positives) end up with a lower value.

As each cluster has a single final numerical value assigned to it corresponding to its overall significance, a thresholding operation can be passed through all final clusters in order to reject those with low confidence.

In order to find a suitable threshold value, an experiment was performed over the full UBEAR test dataset. All final clusters generated in this process were then manually classified as either True Positive or False Positive. Figure 9 shows the distribution of True Positive cluster output values and that of clusters classified as False Positives. After analyzing these distributions, it can be seen that the chosen threshold value of 0.224 most optimally separates it, where a balance can be achieved in rejecting the largest majority of false positive hits, while keeping as many true positives as possible above the threshold.

Note that this process, although similar to traditional Non-Maximum Suppression (NMS), has the added advantage of providing a better filtering mechanism of detections that are likely to be false positives. NMS simply clusters boxes together and keeps the box with highest confidence per cluster, regardless of the distribution of confidence values in the remaining boxes. The proposed method, by comparison, takes into account a weighted distribution of all contributing detections in order to make a more informed decision on the filtering, as this method requires all contributing detections in each clustered set to have a higher confidence value.

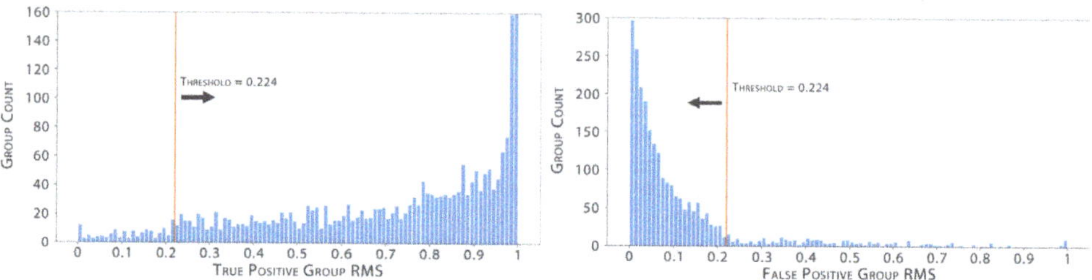

Figure 9. Response of CNN outputs for true positive (TP) and false positive (FP) groups.

A summary of this whole process, starting from the inference, continuing through the grouping, and ending in the thresholding operations, is listed in Algorithm 1:

Algorithm 1 The proposed process including steps for inference, grouping, and applying the threshold.

---

for all $Z \in \{PyramidScales\}$ **do**
    for all $A \in \{L, M, S\}$ **do**
        $O_A^{LE}, O_A^{RE}, O_A^{BG} \leftarrow SharedMap(Image_Z, Network_A)$
    end for
    for all $K \in \{LE, RE, BG\}$ **do**
        $O_{F,Z}^{K} \leftarrow Ensemble(O_S^K, O_M^K, O_L^K)$
    end for
end for
for all $K \in \{LE, RE\}$ **do**
    $G^K \leftarrow Group(O_{F,Z}^K)$
    if $G^K > Threshold$ **then**
        $Keep(G^K)$
    else
        $Discard(G^K)$
    end if
end for

The correct threshold to use should be carefully decided upon depending on the type of data being analyzed. In the case of the AMI database, where images are already prepared as cropped ears, the system detects no False Positives whatsoever, and thus the threshold value decision does not affect the False Positive rate in any way. In this case, a very low (or zero) threshold can be chosen in order to maximize the number of correctly detected ears. This can be seen in the results shown in Figure 10, where the accuracy rate of varying threshold amounts is depicted.

In the case of natural images in non-cooperative environments as with the UBEAR dataset, the effect of false positives is much more important, as can be seen in Figure 10, where small variations in the threshold value lead to a drastic drop in the false positive rate, while not significantly affecting the accuracy of detected ears.

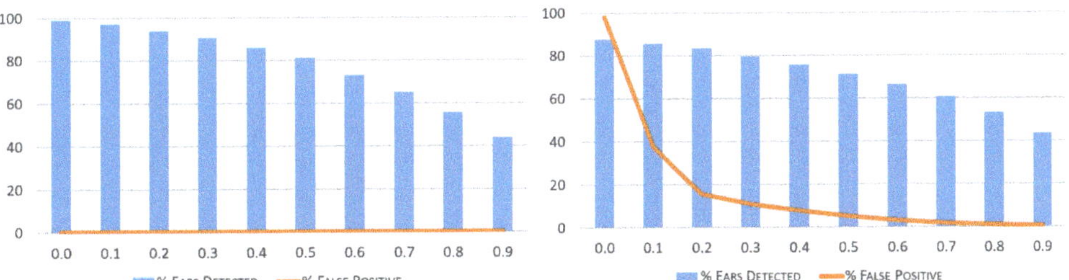

Figure 10. Threshold sensitivity on ear detections: (**Left**) AMI Dataset Detections, (**Right**) UBEAR Dataset Detections.

4. Experiments

4.1. Test Methodology

Multiple experiments were conducted with the various datasets in order to evaluate the system's accuracy in different scenarios. For all tests, the experiment was carried out with the 3-CNN method proposed in this work. To contrast the results, the same tests were also performed with a standard Haar Cascade Classifier trained on similar data as implemented in OpenCV [35], and executed with a similar sliding window configuration while post-processing them with the same window grouping algorithm.

In all cases, the results reported are defined as follows:

- **True Positive:** Detection groups which successfully enclose the bounding box of an ear within the image.
- **False Positive:** Detection groups which mis-classify the side of the ear detected, or which erroneously detect noise in the image that does not correspond to an actual ear.
- **False Negative:** Ears in an image which failed to be detected by the network entirely, or whose final detection group confidence value was below the selected threshold.
- **True Negative:** This value would usually describe the rate at which non-ear noise is successfully ignored by the classifier. However, in the case of full image frames, this would greatly offset the result bias by greatly increasing the overall classification accuracy needlessly. We avoid recording this on purpose such that the results given represent the true nature of correctly classified ears only.

The performance metrics reported for all cases are the precision which measures the exactness of the classifier; the recall which measures its completeness; and the F1 metric which provides a balance between precision and recall, and is therefore a more objective comparison of the performance of two classifiers. Furthermore, the traditional accuracy rate is also reported, in order to provide a basic performance metric.

4.2. Comparison with State of the Art

Due to varied nature of the state of the art in this field, it is very difficult to make a comparative study on performance of our proposed method with all of the existing methods in the literature. In part, this is due to there not being a standard dataset by which all of these algorithms have be benchmarked, but rather every method so far examined in Section 2.1 tends to use their own private data. Similarly, testing existing methods on the same data we use is difficult as most existing implementations remain private and their source code is not readily available for implementation.

Therefore, we can only contribute to Table 1 with our own accuracy results on datasets such as UND and AMI, which are images of similar qualities as the data used in those studies, consisting of ready made images made for this exact purpose. In the case of closeup cropped images such as AMI, our 3-CNN system reaches an accuracy of 99.0% and an F1 metric of 99.50%. On full frame images, such as UND, where localization also plays a part, our system reaches an accuracy of 95.25% and an F1 metric of 97.57%. Full details on these results are found in Section 4.6.

4.3. Video Analysis

Additionally, we also test the detection accuracy on individual video frames. An experiment was carried out with the Video dataset as described in Section 3.1. The purpose of this test is to ensure that both ears can be correctly classified as either left or right, while working with data of variable head poses.

Results of these tests is presented in Table 7, where it can be seen that our system greatly outperforms Haar in this particular task.

Table 7. Results of testing over the Videos dataset.

		Haar				3-CNN			
	Subset Size	Precision (%)	Recall (%)	Accuracy (%)	F1 Score (%)	Precision (%)	Recall (%)	Accuracy (%)	F1 Score (%)
Middle	470	97.60	97.60	95.32	97.60	99.57	99.79	99.36	99.68
Upwards	162	100	69.75	69.75	82.18	95.95	91.03	87.65	93.42
Downwards	284	98.77	57.09	56.69	72.36	94.83	95.19	90.49	95.01
Left Ear	455	97.85	71.21	70.11	82.43	97.07	97.29	94.51	97.18
Right Ear	461	98.53	88.57	87.42	93.29	97.98	96.46	94.58	97.21
Complete Dataset	916	99.05	80.07	79.45	88.55	97.59	96.95	94.68	97.27

The significance of this test is in the ability to continuously detect the same ear on a moving image sequence, regardless of head orientation. The high detection rate ensures that the ear is consistently detected during the majority of each video's duration, except for a few odd frames where detection might fail from time to time. However, a few frames later, the ear is found again and detection continues as normal. This result rate would therefore allow for a tracking mechanism to be successfully implemented in such video streams.

4.4. Image Resolution

Detecting images of subjects at a great distance from the camera is usually problematic. To quantitatively measure the performance of the system in cases where the relative size of the image is very small, various tests were performed on the AMI dataset with the ears previously resized at different scales, ranging from 16×16 up to 96×96. The results of both the combined 3-CNN system as well as that of the individual S, M, and L CNNs are displayed in Figure 11.

This shows that even ears which are found at scales much lower than the networks' input size of 64×64 can still be successfully detected, albeit at a lower rate depending on the actual size.

Figure 11, in particular, explains the dropoff in resolving power at smaller scales. The S CNN is the first one to fail at diminishing scales, as could be expected due to the nature of the data this network analyzes. Meanwhile, the other two CNNs continue to detect with sufficient accuracy at even the smallest scales. Arguably, it could be said that a system without the S scale might do better for this particular purpose, as the dropoff exhibited by the S CNN is the main reason behind the 3-CNN difficulty in detecting smaller sized ears. However, the S CNN has been shown before to be essential for noise differentiation, and as such, this side effect is an acceptable tradeoff.

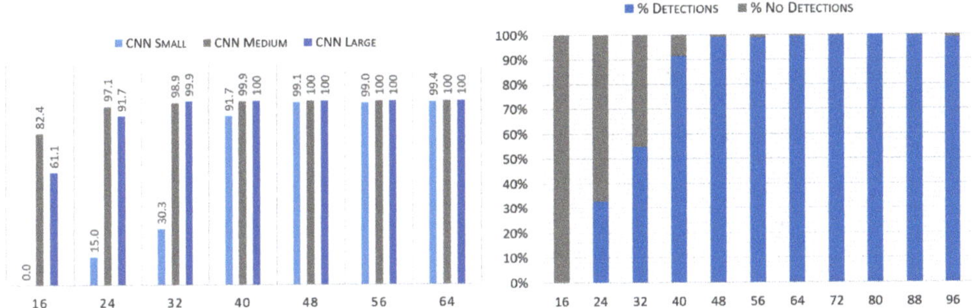

Figure 11. Image resolution and ear size sensitivity: (**Left**) Individual CNNs, (**Right**) 3-CNN System.

4.5. Non-Cooperative Natural Images

Traditional computer vision approaches usually require the ear to be perfectly aligned, or at the very least in the same plane as the photograph projection, thus imposing restrictions that are very restrictive when analyzing real world imagery. Due to the ability of CNNs to learn multiple representations of the same object, and given the pose variety used in the training data, the final trained system is capable of detecting ears at very different angles with respect to the camera.

The UBEAR dataset contains labels for each image which facilitates its partitioning according to the relative pose of the subject in relation to the camera. Tests were run over the full dataset and the results were divided according to the angle of the subject's gaze. These results are depicted in Figure 12 and summarized in Table 8.

The common trend of our 3-CNN outperforming Haar continues to be seen here. However, the real significance behind these results is that Haar, not unlike most traditional computer vision approaches, is highly dependent on viewpoint, and its performance largely drops off as the angle varies from the more normal "Middle" and "Towards" angles. Meanwhile, our 3-CNN system maintains a very similar and stable performance rating regardless of the angle at which the ear is presented.

Further UBEAR labels can be used to split the data into additional folds, such as ear sidedness. As expected, the system works mostly the same for either left or right side ears. The small differences in the results might just be due to a random variation in the images, and not to a real side preference of the classifier.

Finally, we tested the system on images which were marked to have occlusion against those that did not. Occlusion is not a defined label in the UBEAR dataset, therefore, for this study, we manually defined this data fold based on a subjective decision of which images could be considered as occluded. This is because degrees of occlusion can vary from merely a few small strands of hair or a small earring, to very large accessories or full sections of hair covering well over half of the ear. The final *occlusion threshold* decision was made to mark only those ears which had their outline covered at least 25%. This resulted in approximately one third of the images to be marked as occluded.

Not surprisingly, the 3-CNN system performs better when no occlusion is present. However, it is worth noting that even when analyzing occluded ears, the 3-CNN system outperforms Haar when it analyzes clearly visible, and non-occluded ears.

Furthermore, analyzing the literature of existing ear detection systems, such as those described in Table 1, it is obvious that most of the systems which seemingly have very high reported accuracy rates on clearly defined ear images, would drastically fail when the ear is occluded in any way—especially those systems which rely on shape analysis and detection of the tubular or helix properties of an ear.

A final study was performed on gender sensitivity of the detector. The classifiers are not necessarily sensitive to the different shapes of male and female ears. However, a visible disparity can be seen, simply due to the fact that female ears are far more likely to be occluded by longer hair or more prevalent accessories such as large earrings. Thus, gender sensitivity results closely resemble those of occlusion sensitivity.

Figure 13 shows a few selected samples of the 3-CNN and its detection in particularly challenging images, due to either occlusion or extreme viewpoint perspectives.

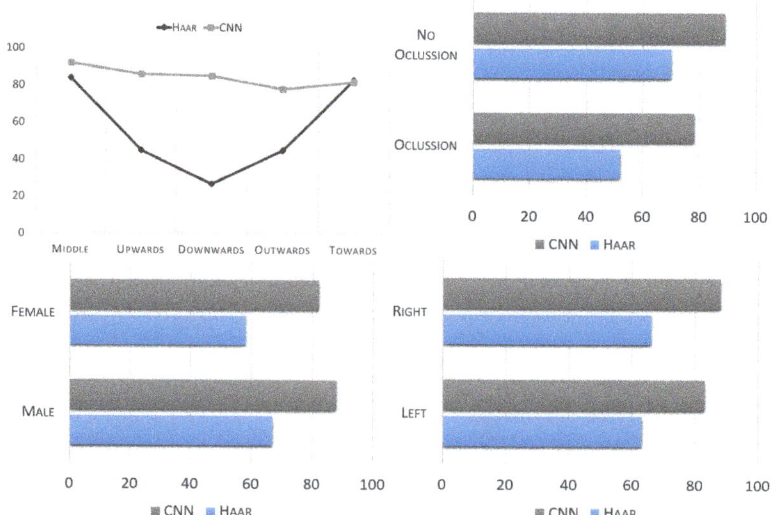

Figure 12. Detection performance of our 3-CNN system vs Haar on the different data folds of the UBEAR dataset: (**Top Left**) Angle Sensitivity, (**Top Right**) Occlusion Sensitivity, (**Bottom Left**) Gender Sensitivity, (**Bottom Right**) Ear Side Sensitivity.

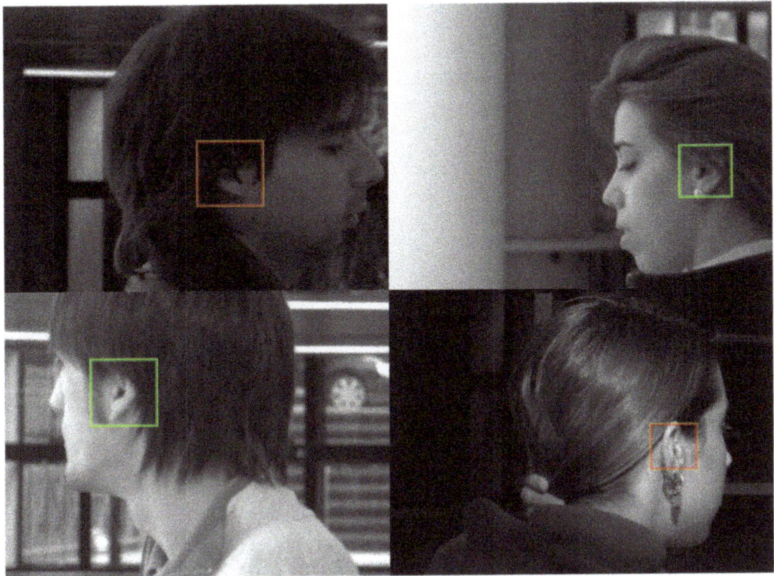

Figure 13. Sample detections on particularly difficult images from the UBEAR dataset, including extreme head orientations and occlusion.

Table 8. Detection performance of our 3-CNN system vs. Haar on the different data folds of the UBEAR dataset.

		Haar				3-CNN			
	Subset Size	Precision (%)	Recall (%)	Accuracy (%)	F1 Score (%)	Precision (%)	Recall (%)	Accuracy (%)	F1 Score (%)
Middle	1392	95.90	74.31	72.03	83.74	89.89	93.92	84.95	91.86
Upwards	813	85.87	30.42	28.97	44.93	87.47	84.94	75.73	86.19
Downwards	784	88.65	16.23	15.90	27.44	85.68	84.57	74.09	85.12
Outwards	789	89.96	30.34	29.35	45.37	85.92	71.26	63.81	77.91
Towards	829	95.10	73.24	70.57	82.75	79.70	84.40	69.47	81.99
Male	3403	94.17	51.58	49.99	66.65	86.10	87.84	76.93	86.96
Female	1204	91.06	42.47	40.77	57.92	86.99	77.83	69.71	82.15
Left Ear	2289	93.49	47.33	45.82	62.84	83.64	83.11	71.48	83.37
Right Ear	2318	93.42	51.07	49.30	66.04	88.97	87.32	78.79	88.14
Occlusion	1491	89.70	36.49	35.03	51.88	85.01	71.63	63.60	77.75
No Occlusion	3116	94.70	55.24	53.59	69.78	86.79	91.53	80.34	89.10
Complete Dataset	4607	93.45	49.22	47.58	64.48	86.31	85.23	75.08	85.77

4.6. Summary

To conclude, Table 9 lists a summary of all total results across all four datasets while comparing our 3-CNN system with the well known Haar Cascade Classifier algorithm.

Table 9. Summary of the total results over all four datasets contrasting the Haar and 3-CNN algorithms.

Dataset	Algorithm		Positive	Negative	Precision (%)	Recall (%)	Accuracy (%)	F1 Score (%)
UND [32,33]	3-CNN	Positive	461	20	95.84	99.35	95.25	97.57
		Negative	3	0				
	Haar	Positive	270	7	97.47	58.44	57.57	73.07
		Negative	192	0				
Videos	3-CNN	Positive	890	22	97.59	96.95	94.68	97.27
		Negative	28	0				
	Haar	Positive	727	7	99.05	80.07	77.47	87.31
		Negative	181	0				
AMI [31]	3-CNN	Positive	693	0	100.00	99.00	99.00	99.50
		Negative	7	0				
	Haar	Positive	382	7	98.20	55.12	54.57	70.61
		Negative	311	0				
UBEAR [34]	3-CNN	Positive	3814	605	86.31	85.23	75.08	85.77
		Negative	661	0				
	Haar	Positive	2227	156	93.45	49.22	47.58	64.48
		Negative	2298	0				

As can be seen, the CNN based system always outperforms the Haar algorithms in all sets, by an amount ranging between 10% to 29% in the F1 metric. This is particularly so in the UBEAR dataset, since the Haar classifier is incapable of modelling the higher variety of internal representations required to properly classify images in that dataset.

Figure 14 shows a summary of these results. It is important to remark that that our proposed system has stable performance figures across the first three datasets, all of which consist of perfect purpose-made ear photography. The results only slightly drop when presented with natural images due to the challenges already described. This is in contrast to the Haar classifier, which has wildly disparate results, demonstrating the large dependency of this system on the particular conditions of one dataset or another.

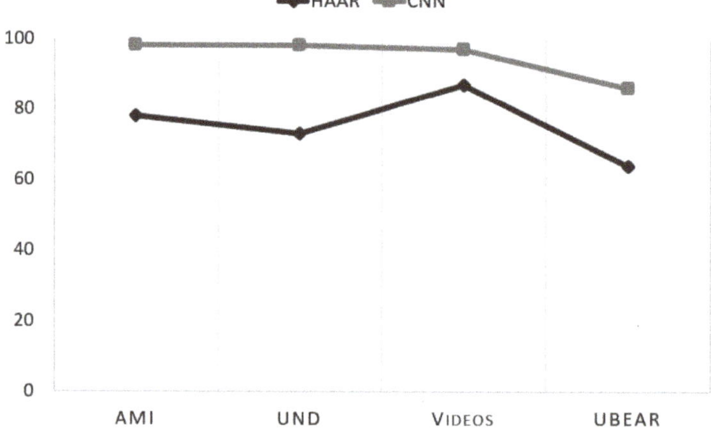

Figure 14. Results of our 3-CNN system compared to the Haar classifier over the various test datasets.

5. Conclusions

We propose a new technique based on CNNs to carry out ear detection on natural images. As opposed to traditional computer vision approaches that are based on hand-crafted features, Convolutional neural networks perform image and shape perception, which is far more robust against variable perspective, occlusion and illumination conditions. These difficult conditions are very common in natural images, compared to synthetic photographs taken in strictly controlled photographic and illumination conditions.

All previously proposed systems usually fail in one important way or another. Some require the ear to be properly aligned. Others require the full ear to be visible. Most commonly, they are highly sensitive to illumination and require images shot in the exact same conditions as the training data, or they may even fail when the images are not fully in focus or when the relative size of the ear in the image is not sufficiently large.

Up to now, we have not seen a robust all-encompassing system capable of detecting ears under all possible conditions in natural images, and we are glad to introduce this new alternative. Granted, our system still has some important failures which we must address in future versions of the system, primarily to decrease the false positive rate, which would allow decreasing the threshold and thus improve the overall performance. However, the results so far are very encouraging, and having such a robust detector is the first important step towards building an ear recognition system, something which obviously is a future line of research to be conducted presently.

Further future lines of research include the implementation of this system in an even more optimized manner in order to deploy it on low power mobile or embedded devices for practical biometric applications.

Finally, it is important to note that although this work was aimed mainly towards ear detection, it presents an end-to-end object recognition framework which can be adapted very similarly to other computer vision tasks requiring a comparable type of classification executed over natural imagery for real-time detection and tracking. Convolutional neural networks have been shown time and time again to be extremely powerful image classifiers, especially when they are used as ensemble systems, and this work has presented one more way in which they can be applied to this kind of task.

Author Contributions: Conceptualization, W.R. and P.G.; methodology, W.R.; software, W.R.; resources, W.R.; investigation, W.R.; validation, W.R. and P.G.; formal analysis, P.G.; data curation, P.G.; writing—original draft preparation, W.R. and P.G.; writing—review and editing, W.R. and M.G.; visualization, P.G.; supervision, M.G.; project administration, M.G.

Funding: This research received no external funding.

Conflicts of Interest: The authors declare no conflict of interest.

References

1. Galdámez, P.L.; Arrieta, A.G.; Ramón, M.R. Ear Recognition Using a Hybrid Approach Based on Neural Networks. In Proceedings of the 17th International Conference on Information Fusion (FUSION), Salamanca, Spain, 7–10 July 2014; pp. 1–6.

2. Krizhevsky, A.; Sutskever, I.; Hinton, G.E. ImageNet Classification with Deep Convolutional Neural Networks. In *Advances in Neural Information Processing Systems 25*; Curran Associates, Inc.: Red Hook, NY, USA, 2012; pp. 1097–1105.

3. LeCun, Y.; Bottou, L.; Bengio, Y.; Haffner, P. Gradient-Based Learning Applied to Document Recognition. *Proc. IEEE* **1998**, *86*, 2278–2324 [CrossRef]

4. Ciresan, D.; Meier, U.; Masci, J.; Schmidhuber, J. Multi-Column Deep Neural Network for Traffic Sign Classification. *Neural Netw.* **2012**, *32*, 333–338. [CrossRef]

5. Ciresan, D.; Meier, U.; Gambardella, L.; Schmidhuber, J. Convolutional Neural Network Committees for Handwritten Character Classification. In Proceedings of the 11th International Conference on Document Analysis and Recognition (ICDAR), Beijing, China, 18–21 September 2011.

6. Burge, M.; Burger, W. Ear Biometrics in Computer Vision. In Proceedings of the 15th International Conference on Pattern Recognition (ICPR-2000), Barcelona, Spain, 3–7 September 2000.

7. Hurley, D.; Nixon, M.; Carter, J. Force Field Energy Functionals for Image Feature Extraction. *Image Vis. Comput.* **2002**, *20*, 311–317. [CrossRef]

8. Yan, P.; Bowyer, K.W. Empirical Evaluation of Advanced Ear Biometrics. In Proceedings of the International Conference on Computer Vision and Pattern Recognition-Workshop, San Diego, CA, USA, 21–23 September 2005; Volume 3.

9. Chen, H.; Bhanu, B. Shape Model-Based 3D Ear Detection from Side Face Range Images. In Proceedings of the IEEE Computer Society Conference on Computer Vision and Pattern Recognition-Workshops, San Diego, CA, USA, 21–23 September 2005.

10. Chen, H.; Bhanu, B. Contour Matching for 3D Ear Recognition. In Proceedings of the Seventh IEEE Workshop on Applications of Computer Vision, San Diego, CA, USA, 21–23 September 2005.

11. Chen, H.; Bhanu, B. Human Ear Recognition in 3D. *IEEE Trans. Pattern Anal. Mach. Intell.* **2007**, *29*, 718–737. [CrossRef] [PubMed]

12. Attarchi, S.; Faez, K.; Rafiei, A. A New Segmentation Approach for Ear Recognition. In *Advanced Concepts for Intelligent Vision Systems*; Springer: Berlin/Heidelberg, Germany, 2008.

13. Cummings, A.; Nixon, M.; Carter, J. A Novel Ray Analogy for Enrollment of Ear Biometrics. In Proceedings of the Fourth IEEE International Conference on Biometrics: Theory, Applications and Systems (BTAS), Washington, DC, USA, 27–29 September 2010.

14. Amioy, K.; Madasu, H.; Mohit, K.; Gupta, H. Automatic Ear Detection for Online Biometric Applications. In Proceedings of the Third National Conference on Computer Vision, Pattern Recognition, Image Processing and Graphics, Karnataka, India, 15–17 December 2011.

15. Lankton, S.; Tannenbaum, A. Localizing Region-Based Active Contours. *IEEE Trans. Image Proc.* **2008**, *17*, 2029–2039. [CrossRef] [PubMed]

16. Pflug, A.; Busch, C. Ear Biometrics: A Survey of Detection, Feature Extraction and Recognition Methods. *IET Biom.* **2012**, *1*, 114–129. [CrossRef]

17. Abaza, A.; Hebert, C.; Harrison, M. Fast Learning Ear Detection for Real-time Surveillance. In Proceedings of the Fourth IEEE International Conference on Biometrics: Theory Applications and Systems, Washington, DC, USA, 27–29 September 2010.

18. Ansari, S.; Gupta, P. Localization of Ear Using Outer Helix Curve of the Ear. In Proceedings of the International Conference on Computing: Theory and Applications, Kolkata, India, 5–7 March 2007.

19. Alvarez, L.; Gonzalez, E.; Mazorra, L. Fitting Ear Contour Using an Ovoid Model. In Proceedings of the 39th Annual International Carnahan Conference on Security Technology, Las Palmas, Spain, 11–14 October 2005.

20. Arbab-Zavar, B.; Nixon, M. On Shape-Mediated Enrollment in Ear Biometrics. In *Advances in Visual Computing*; Springer: Berlin/Heidelberg, Germany, 2007.

21. Arbab-Zavar, B.; Nixon, M. Robust Log-Gabor Filter for Ear Biometrics. In Proceedings of the International Conference on Pattern Recognition, Tampa, FL, USA, 8–11 December 2008.

22. Islam, S.; Bennamoun, M.; Davies, R. Fast and Fully Automatic Ear Detection Using Cascaded Adaboost. In Proceedings of the Applications of Computer Vision, Copper Mountain, CO, USA, 7–9 January 2008.

23. Jeges, E.; Mt, L. Model-Based Human Ear Localization and Feature Extraction. *Int. J. Intell. Comput. Med. Sci. Image Process.* **2007**, *1*, 101–112. [CrossRef]

24. Liu, H.; Liu, D. Improving adaboost ear detection with skin-color model and multi-template matching. In Proceedings of the 3rd IEEE International Conference on Computer Science and Information Technology, Chengdu, China, 9–11 July 2010.

25. Prakash, S.; Gupta, P. An Efficient Ear Localization Technique. *Image Vis. Comput.* **2012**, *30*, 38–50. [CrossRef]

26. Shih, H.; Ho, C.; Chang, H.; Wu, C.S. Ear Detection Based on Arc-Masking Extraction and Adaboost Polling Verification. In Proceedings of the Fifth International Conference on Intelligent Information Hiding and Multimedia Signal Processing, Kyoto, Japan, 12–14 September 2009.

27. Yan, P.; Bowyer, K. Biometric Recognition Using 3D Ear Shape. *IEEE Trans. Pattern Anal. Mach. Intell.* **2007**. [CrossRef] [PubMed]

28. Yuan, L.; Mu, Z.C. Ear Detection Based on Skin-Color and Contour Information. In Proceedings of the International Conference on Machine Learning and Cybernetics, Hong Kong, China, 19–22 August 2007.

29. Nasse, F.; Thurau, C.; Fink, G. Face Detection Using GPU-Based Convolutional Neural Networks. *Comput. Anal. Images Pattern* **2009**, *5702*, 83–90.

30. Raveane, W.; González Arrieta, M. Shared Map Convolutional Neural Networks for Real-Time Mobile Image Recognition. In *Distributed Computing and Artificial Intelligence, 11th International Conference*; Springer: Cham, Germany, 2014; Volume 290, pp. 485–492.

31. Gonzalez, E.; Alvarez, L.; Mazorra, L. AMI: Ear Database. Available online: http://www.ctim.es/research_works/ami_ear_database/ (accessed on 5 June 2019).

32. Flynn, P.J.; Bowyer, K.W.; Phillips, P.J. Assessment of Time Dependency in Face Recognition: An Initial Study. In *Audioand Video-Based Biometric Person Authentication*; Springer: Berlin/Heidelberg, Germany, 2003.

33. Chang, K.; Bowyer, K.W.; Sarkar, S.; Victor, B. Comparison and Combination of Ear and Face Images in Appearance-Based Biometrics. *IEEE Trans. Pattern Anal. Mach. Intell.* **2003**, *25*, 1160–1165. [CrossRef]

34. Raposo, R.; Hoyle, E.; Peixinho, A.; Proença, H. UBEAR: A Dataset of Ear Images Captured On-The-Move in Uncontrolled Conditions. In Proceedings of the IEEE Workshop on Computational Intelligence in Biometrics and Identity Management, Paris, France, 11–15 April 2011.

35. Bradski, G.; Kaehler, A. OpenCV. In *Dr. Dobb's Journal of Software Tools*; Consulting Prof. Stanford University: Stanford, CA, USA, 2000.

Article

An Accurate Clinical Implication Assessment for Diabetes Mellitus Prevalence Based on a Study from Nigeria

Muhammad Noman Sohail [1,*]**, Ren Jiadong** [1]**, Musa Uba Muhammad** [1]**,
Sohaib Tahir Chauhdary** [2]**, Jehangir Arshad** [2] **and Antony John Verghese** [3]

[1] Department of Information Sciences and Technology, Yanshan University, Qinhuangdao 066000, China;
jdren@ysu.edu.cn (R.J.); musaubamuhammad@stumail.ysu.edu.cn (M.U.M.)

[2] Department of Electrical and Computer Engineering, COMSATS University Islamabad, Islamabad 43600,
Pakistan; sohaibchauhdary@hotmail.com (S.T.C.); jehangir@cuisahiwal.edu.pk (J.A.)

[3] Department of Management, American Hotel and Lodging Association, New York, NY 10006, USA;
antomatter@gmail.com

* Correspondence: mn.sohail@stumail.ysu.edu.cn; Tel.: +86-1503-237-0085

Received: 9 April 2019; Accepted: 10 May 2019; Published: 15 May 2019

Abstract: The increasing rate of diabetes is found across the planet. Therefore, the diagnosis of pre-diabetes and diabetes is important in populations with extreme diabetes risk. In this study, a machine learning technique was implemented over a data mining platform by employing Rule classifiers (PART and Decision table) to measure the accuracy and logistic regression on the classification results for forecasting the prevalence in diabetes mellitus patients suffering simultaneously from other chronic disease symptoms. The real-life data was collected in Nigeria between December 2017 and February 2019 by applying ten non-intrusive and easily available clinical variables. The results disclosed that the Rule classifiers achieved a mean accuracy of 98.75%. The error rate, precision, recall, F-measure, and Matthew's correlation coefficient MCC were 0.02%, 0.98%, 0.98%, 0.98%, and 0.97%, respectively. The forecast decision, achieved by employing a set of 23 decision rules (DR), indicates that age, gender, glucose level, and body mass are fundamental reasons for diabetes, followed by work stress, diet, family diabetes history, physical exercise, and cardiovascular stroke history. The study validated that the proposed set of DR is practical for quick screening of diabetes mellitus patients at the initial stage without intrusive medical tests and was found to be effective in the initial diagnosis of diabetes.

Keywords: data mining; cluster; clinical implications; diabetes; epidemiology; forecast; PART; Decision table; Weka; real-life patients; regression; machine learning

---

1. Introduction

Diabetes mellitus (DM) is an exponentially growing disease across the developing countries of the 21st century. Diabetes mellitus has now become a worldwide challenge and identified as the risk factor of other chronic diseases such as hyperosmolar, diabetic ketoacidosis, and hyperglycemia and, in extreme cases, death. Furthermore, diabetes also causes long-term complications, for instance, cardiovascular disease, heart stroke, kidney failure, chronic ulcers, blindness, damage to the eyes, and many more [1]. Williams wrote in his book "Williams textbook of Endocrinology" [2] that around 385 million people were affected with diabetes in 2013. If Diabetes mellitus is left untreated, this figure can get higher; it can even lead to death. Around 425 million people had diabetes in the world by the survey report of the International Diabetes Federation (IDF) in 2015 [3]. Also, the report indicates that 382 million people around the globe are affected by diabetes in developing countries alone and Africa has 4.9% from this ratio.

By the World Health Organization (WHO) [4], 321,100 deaths occurred in the African region due to diabetes, out of which 79% of the population was under the age of 60; this is the maximum number in any region of the world. The ratio of diabetes mellitus patients in rural and urban areas of Nigeria varies from 0.67% to 12%, and this ratio has been estimated to more than double over the past two decades [5]. According to the IDF report, the ratio of undiagnosed diabetic people in sub-Saharan Africa (SSA) is estimated at 87%, out of which 8.7% in the male and 8.9% in the female population of Nigeria. It is due to the lack of information and government resources [6]. In addition, the American Diabetes Association (ADA) estimates that the prevalence was estimated in Nigeria as 20.01% in both the male and female population [7]. Compared with the world population, the Nigerian health organizations pointed out that the diabetes prevalence was 4.7% in 2010 and it was projected to be 5.8% by 2030 and even exceed 10% by 2040 [8]. However, this estimate comes from rural areas, and it is expected to be more in urban slums.

In this study, the fundamental objective was to develop a quick and accurate prediction assessment scheme by using easily observable clinical features to identify patients with a high risk of diabetes. For this purpose, the machine learning Rule classifiers (projective adaptive response theory (PART) and Decision table) were used on the Weka 3.9.2 platform for acquiring accuracy in classification assumptions. Afterward, the logistic regression (LR) was utilized on the classification results to predict and forecast patients with a high risk of diabetes. This research can be applied to diabetes mellitus patients who cannot afford the expenses of the medical laboratory and specifically those in remote areas or villages with low socioeconomic status and excessive epidemiological risk.

Correspondingly, the remaining paper is structured as follows: Section 2 explains the material and methodology after the background description, Section 3 reviews the results, Section 4 discusses the results and limitations, and Section 5 concludes the findings.

Background

Numerous authors work to develop appropriate disease prediction algorithms. For instance, Lélis et al. applied seven classification techniques in a Brazilian investigation to make a diagnosis of meningococcal meningitis and verified that the model is affordable and accurate [9]. Susanne et al. proposed a mathematical model to forecast the prevalence of diabetes by using attributes of sex, age, risk factor status, and T2DM (type 2 diabetes mellitus) status and found T2DM prevalence is projected to increase by 43%, and the incidence is projected to increase 147% by 2050 in Qatar [10]. Choi et al. applied support vector machine (SVM) and artificial neural network (ANN) to screen the pre-diabetes of 9251 individuals and performed a systematic assessment of the models using external and internal cross-validation and concluded that the results of the SVM method are better than the ANN [11]. Amir et al. proposed a time series prediction model for the diagnosis of diabetes patients [12]. In addition, Olivera et al. utilized machine learning algorithms from ELSA-Brazil and identified individuals with the highest risk of undiagnosed diabetes from readily available clinical data [13]. Sohail et al. performed the classification results on Weka by machine learning by utilizing the dataset of different diseases and concluded the accuracy ratio of the decision tree (86%), the Bayesian network (90%), the naïve Bayesian (76%), the fuzzy cognitive map (94%), and K-nearest neighbor (KNN) (94%) [14]. Parampreet et al. applied a cloud-based framework with the help of sensor devices to initially screen patients for the prediction of diabetes [15]. Further, Hassan et al. proposed a unified machine-learning framework for diabetes predications in big data [16]. There is considerable interest in determining how different classification techniques from machine learning can be utilized as disease prediction tools [17–21]. These tools have been used to diagnose diabetes [22], glaucoma [23], meningitis [24], coronary artery disease [25], asthma [26], cancer [27], hypertension [28], heart arrhythmia [29], tuberculosis [30], and other diseases [31,32].

2. Material and Methods

2.1. Ethical Consents

The study was approved by the Natural Science Foundation of China Hebei province, the Yanshan University ethics committee, and all experiments and simulation procedures conformed to the Declaration of Helsinki. All participants provided written informed consent after having all procedures explained to them both verbally and in writing.

2.2. Model Framework

Figure 1 shows the assessment framework used in this study for diabetes patient screening. The assessments were performed in a total of six steps. Initially, the real-life diabetes mellitus data were acquired and preprocessed for selection of appropriate attributes. Afterward, this data was utilized for evaluation and assessment. Secondly, the updated plugins of two machine learning Rule classifiers (PART and Decision table) were used on Weka version 3.9.2 "data mining platform" for classification measurements and Rule assessment [33]. In addition, the logistic regression method was utilized on the results of the machine learning classifiers to forecast the rule assessment.

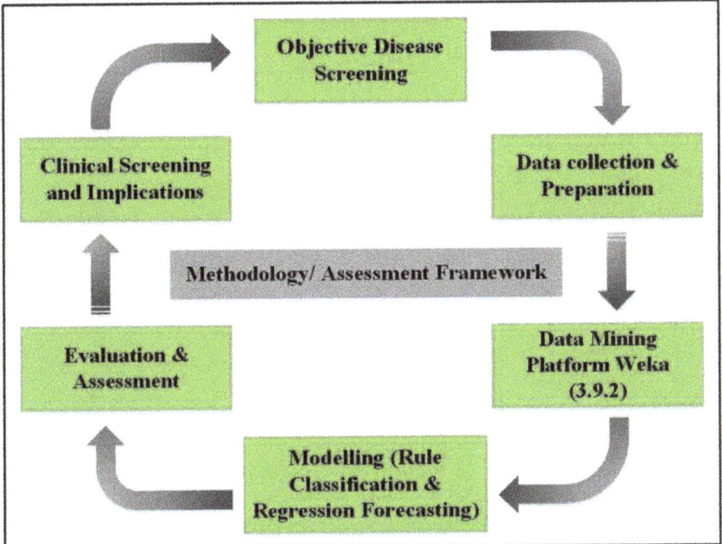

Figure 1. Assessment framework used in this study for clinical implication screening.

2.3. Data Collection and Explanation

The real-life diabetes mellitus data of 1257 patients from December 2017 to February 2019 were acquired from four main hospitals across Nigeria and carefully examined. Figure 2 demonstrate the collection flow of data gathered from four principal hospitals in Nigeria namely Abdullahi Wase Specialist Hospital (22.75%), Ajingi General Hospital (22.04%), Federal Medical Center Birnin-Kudu (26.81%), and Gaya General Hospital (28.40%) located in the northwestern region of Nigeria. The data were collected through questionnaires, verbal interviews, and by consultation of the medical specialist after the ethics committee of the institute where the research was carried out approved the study protocols. The data collection flow of diabetes patients from the mentioned hospitals is shown in Figure 2, and the number of patients in each hospital is shown in Figure 3.

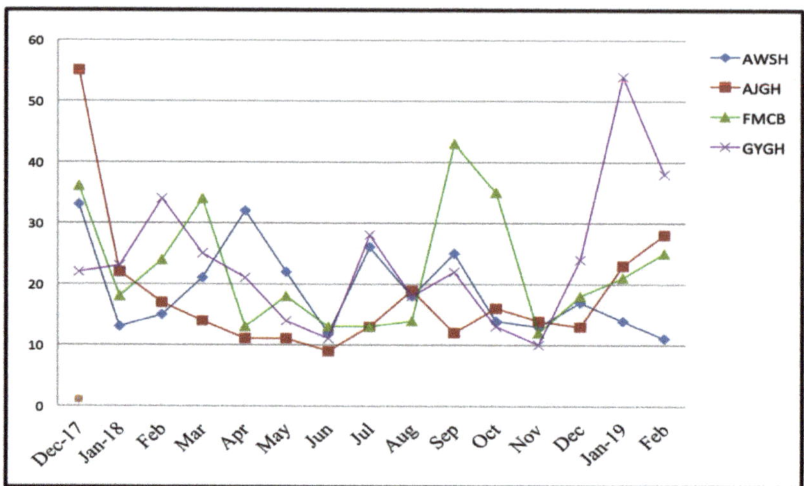

Figure 2. The data collection flow of diabetes patients from the four hospitals.

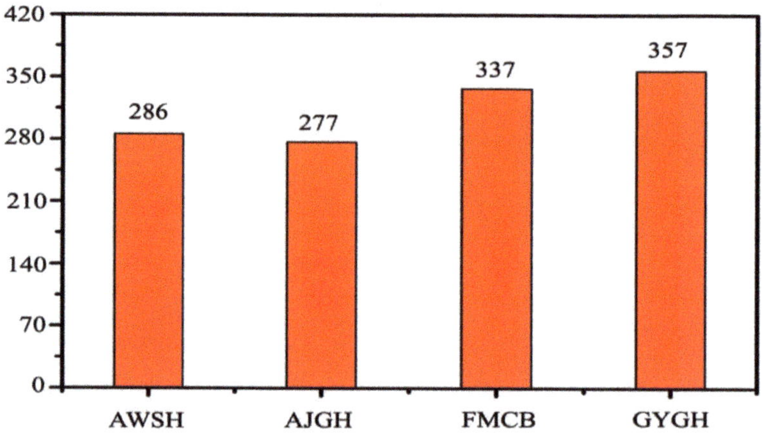

Figure 3. Total number of diabetes patients recorded in the four hospitals.

2.4. Attributes Selection

In our prediction assessment of diabetes mellitus prevalence, the data of 10 easily available attributes/variables, namely age, gender, GLU (glucose level of the patient), BMI (body mass index of the patient), HYP (hypertension status), HCD (history of cardiovascular disease), FDH (family history of diabetes), PEX (physical exercise), STW (work stress status), and DIT (diet of the patient, healthy and unhealthy). Out of 1257 records, 587 patient records were missing values in the body mass index, glucose level, hypertension, cardiovascular disease, work stress status, family diabetes experience, physical exercises, and diet lifestyles. Moreover, 389 records were removed from assessment dataset because of missing values in pre-diabetes status. Therefore, 281 records with 10 variables were used in the prediction analysis.

2.5. Attribute Parameters

The 10 features selected in this study were characterized as follows. Age and gender represented demographic characteristics. A patient's glucose level (mmol/L) has a relation with age and diet.

Family history of diabetes was defined as any family member previously diagnosed by a physician as diabetic or pre-diabetic (Yes = 1, No = 0). BMI was calculated as body weight divided by the square of height in meters and BMI ≥ 25 was defined as overweight. History of cardiovascular disease or stroke was defined as the patient having been previously diagnosed with coronary heart disease or stroke by a surgeon (Yes = 1, No = 0). Physical exercise indicated whether the patient engaged in exercise (Yes = 1, No = 0). Work stress was measured to the patient's subjective impression (Yes = 1, No = 0). Diet was measured as a balanced and unbalanced diet (Yes = 1, No = 0). HYP was defined in three ways: first, a systolic BP (blood pressure) ≥140 mmHg; second, medication for BP control; and third, diastolic BP ≥ 90 mmHg.

2.6. Data Mining Platform

Waikato Environment for Knowledge Analysis (Weka/v 3.9.2) was used for the preprocessing and classification assessment of diabetes mellitus by updated plugins of the Kmean clustering algorithm to assign the class to the dataset of 10 variables for testing as positive and negative status (positive mean diabetes and negative mean normal status) [34]. The positive patients were declared as high in diabetes status after assessment and negative as normal for the initial screening by proper forecast assessment. The advantage of using Weka is the avoidance of overfitting and unnecessary complexity.

In addition, Rule algorithms (PART and Decision table) were adopted for accurate measurements. Moreover, the logistic regression was utilized on the assessment of classification to forecast diabetes prevalence for clinical implications.

After data preprocessing, the final dataset included 281 patient records with males and females and 11 attributes. The population sampling included patients with diabetes mellitus status Type 1 (non-insulin dependent), Type 2 (insulin-dependent), and gestational diabetes. The 11 attributes included 10 as input attributes and the one as the target attribute. The target attribute consisted of two classes: one class obtained the diagnosis of diabetes tested positive and the second class was tested negative by the Kmean finding within the clusters that are more related to each other at the significance level of 0.05 [35].

Kmean is a typical distance-based cluster algorithm and its distance is measured on similarities. The process steps of the Kmean are to measure the distance between each object and the centers of the cluster by Equations (1)–(3), as follows:

$$S_i^{(t)} = \left\{ \forall j, 1 \text{A} j \text{A} k X_p : \|X_p - m_i^t\|^2 \leq \|X_p - m_j^t\|^2 \forall j, 1 \leq j \leq k \right\} \forall j, 1 \text{A} j \text{A} k, \tag{1}$$

$$m_i^{t+1} = \frac{1}{\left|S_i^{(t)}\right|} \sum_{x_j \in S_i^{(t)}} X_j, \tag{2}$$

$$J = \sum_{j=1}^{k} \sum_{i=1}^{n} \| x_i^{(j)} - c_j \|^2, \tag{3}$$

where n is the number of data points in the *i* clusters, k is the number of cluster centers, and $\| x_i^{(j)} - c_j \|$ represents the Euclidean distance between $x_i^{(j)}$ and c_j. In addition, the Kmean clustering algorithm is composed of the following steps.

(i)   Place the K points into the considerable space as represented by the objects that are being clustered, which indicate the initial group of centroids.

(ii)   Properly assign each object to the group that undoubtedly possesses the most adjacent centroid.

(iii)   After assigning all objects, recalculate the prominent position of the K centroid.

(iv)   Repeat the second and third step until the centroids are not able to shift significantly more. This efficiently produces the possible separation of group objects, which can accurately calculate the matrix to be minimized by Equation (4).

$$argmin_{c_j \in C} \ dist(c_i, x)^2. \tag{4}$$

2.7. Rules Classification

The machine learning algorithms PART and Decision table were utilized for the classification of the dataset with a 10-fold cross-validation assessment. PART classifications are projected as discrete rules to conquer the rule methods of any dataset and generate a rule set for a better understanding of the decision list. In addition, PART works with a combination of C4.5 and Ripper [36]. The paramount leaf in the rules assessment was generated by the fractional C4.5 decision tree repetitions. It compares the data to the rules of each list, and vice versa, and assigns the items accordingly.

The decision table summarizes the testing dataset and compares it with the training dataset generated. In addition, it classifies the unknown dataset samples by the Wrapper method, which helps to reduce the unknown values and produce better results with higher accuracy and minimal error rates [37]. The first attribute in the rule tree is the most informative node, which is measured by Equations (5) and (6):

$$I_A = E(D) - \sum_{i=1}^{k} \frac{|D_i|}{|D|} E(D_i), \tag{5}$$

$$E(X) = -\sum_{i=1}^{m} \frac{count(c_i, x)}{|x|} \cdot log \frac{count(c_i, x)}{|x|}. \tag{6}$$

The parameter selected for the PART classifier was 100 as the batch size with false in binary splits by a confidence interval of 0.25%. The number of objects was set as 2, decimal number places as 2, fold number as 3, error pruning as false, and seeds value as 1. In addition, the parameters for the Decision table were 100 as the batch size with a cross value was 1 and the number of decimal places was 2 with the best first in search results.

2.8. Kappa Statistics

Kappa statistics have the consistency of frequent testing, which provides extended facts about data collection in the research that is correct for variable measurements. It compares the model results with the randomly generated classification. We adopted kappa stats measures based on values between 0 and 1 as in Equations (7)–(9) where the value 0 is invalid and 1 is the expected effect of the assessment. Furthermore, kappa stats indicate the consistency of assessment.

$$K = [P(A) - P(E)]/[1 - P(E)] \tag{7}$$

$$P(A) = [(TP + TN)/N] \tag{8}$$

$$P(E) = [(TP + FN) * (TP + FP) * (TN + FN)/N^2 \tag{9}$$

2.9. Logistic Regression Forecasting

Logistic regression was implemented on the classification outcomes with the primary objective to define the initial screening for disease diagnosis and prediction [38]. In most cases, the variables of the logistic regression work to solve the two-way binary classifications. It predicts the continuous values to maintain the sensitivity in the numbers field where the values are 0 and 1. The value 1 is assigned only if the value is greater than the threshold (value > threshold); otherwise, it will be 0. Hence, the range of output works in the logistic regression is between 0 and 1 with the addition of the sigmoid function layers measured by Equations (10)–(13):

$$P = \alpha + \beta_1 X_1 + \beta_2 X_2 + \ldots + \beta_m X_m, \tag{10}$$

$$\sigma(x) \frac{1}{1 + e^{-x}} \in [0, 1], \tag{11}$$

$$Pr(Y = +1|X) \sim \beta.X, \tag{12}$$

$$Pr(Y = -1|X) = 1 - Pr(Y = +1|X). \tag{13}$$

It consists of a positive and a negative group of values. The variable X will be assigned to the β coefficient values, which represent the weight. Y indicates the patients with diabetes. The variations between the values X and Y occur on the basis of weight.

The parameters selected for the logistic regression forecast was 1 for a number of time units. The confidence interval was set at 0.95%. The M5 method was chosen for attribute selection with a batch size of 100, and the ridge was set as 1.0 E-8. After accurately setting up, it is easy to predict the outcome of positive or negative. The sigmoid function $\sigma(x)$ proposition is described as follows:

Proposition 1. A function $f: (0,1) \rightarrow R$ is absolutely a monotone on $(0,1)$ if and only if it possesses a power series expansion with non-negative coefficients, converging for $0 < x < 1$.

Proof. If (f) function is completely monotone in $(0,1)$, then the power series expansion of (f) function in $(0,1)$ has to be alternating because $(-1)^k f^k \geq 0$. On the other hand, consider an alternating power series of function $f(x)$ converging for all $0 < x < 1$ and its derivatives by Equations (14)–(16):

$$f(x) = a_0 - a_1 x + a_2 x^2 - a_3 x^3 \ldots a_i \geq (0 < x < 1), \tag{14}$$

$$(-1)f^1(x) = a_1 - 2a_2 x + 3a_3 x^3 + \ldots, \tag{15}$$

$$f^2(x) = 2a_2 - 6a_3 x + \ldots \tag{16}$$

□

3. Results

A total of 281 diabetes patients were evaluated; 121 (43.06%) were male and 160 (56.93%) were female. Among the 281 records, 256 (91.10%) were not dependent on insulin (Type 1), 14 (4.98%) were Gestational, and 11 (3.91%) were insulin dependent (Type 2). Initially, the dataset was divided into a 20:80 ratio for conducting training and testing. After training the machine, a 10-fold cross-validation technique was implemented on an experimental platform of Weka for better assessment of the classification. The dataset was divided into 10 samples. Each sample was utilized as validation data from the retention process, while the remaining nine samples served as the training data. This process was performed 10 times. The advantage of this process is the reduction in the error ratio and bias correlation by random sampling.

3.1. Measurements

Initially, the PART rule classifier was tested on the dataset to measure the classification accuracy with the seed of random numbers selected for XVal. The percentage was 1, the confidence factor was 0.25%, the minimum number of objects was 2, and the number of folds was set to 3. After loop tests, the average accuracy of the final result was 99.28%. Secondly, the same measurement was tested on the Decision table rule classifier. The final result with an average accuracy of 98.22% was obtained in 0.77 s. The subset value was 99.60%, and the average error was 0.03%. By employing the rule classification (PART and Decision table), good predictive rules were obtained for the patient's care. The outcomes in the initial phase were the most appropriate with a mean accuracy of 98.75%; the error rate remained at 0.02%.

The results obtained for the classification accuracy are presented in Table 1 along with the attribute details and the clustering instances for the classification. It is comprised of three sections. The first section discusses the details of the properties used for the Weka platform for assessment, with 281 patients describing their age limits by classification type and improving the evaluation of positive and negative tested weights. Additionally, it provides accurate information and average classification accuracy for PART and Decision table rule classifiers, including kappa statistics, mean error, true positive rate, false positive rate, accuracy, recall rate, F-measure, Matthew's correlation curve (MCC)., Receiver operating characteristics (ROC), Precision recall curve (PRC) area ratios, and the time it takes for a prediction analysis [39,40].

Table 1. Diabetes type and the number of patients classified for the Rule assessment.

Diabetes Type	Patients (N = 281)	Age	Weight	"0" Missing Values	Attributes	Class	
						T_N	T_P
NID	256		256.0				
GTD	14	>10 <87	14.0		11	87	194
IND	11		11.0				

Classification	PART Rule %	Decision Table Rule %

Total number of diabetes mellitus patients from age >10 and <87 (*N* = 281)

- From age ≤20 = 2 patients
- From age >20 and ≤40 = 58 patients
- From age >40 and ≤60 = 144 patients
- From age >60 and ≤80 = 76 patients
- From age >80 = 1 patient

Classification	PART Rule %	Decision Table Rule %
Accuracy	99.28	98.22
Kappa statistics	0.98	0.96
Mean absolute error	0.01	0.03
True positive rate	0.99	0.98
False positive rate	0.01	0.01
Precision	0.99	0.98
Recall	0.99	0.98
F-Measure	0.99	0.98
MCC	0.98	0.96
ROC area	0.99	0.99
PRC area	0.99	0.99
Time taken to build the model	0.10 s	0.77 s
Average accuracy	99.28	98.22
Mean average accuracy	98.75%	

Values	Counts (N′ = 281)	Ratio	Cluster by Class		Cluster by Diabetes Type		
			T_N	T_P	NID	GTD	IND
0	138	49%	47	91	128	7	3
1	143	51%	40	103	128	7	8 [1]

[1] NID = not insulin dependent; GTD = gestational diabetes patients; IND = insulin dependent; MCC = Matthew's correlation curve; ROC = Receiver operating characteristics; PRC = Precision recall curve; N = number of patients; ≥greater than; ≤less than; % = percentage value; T_N = tested negative; T_P = tested positive; Values = two clusters 0 and 1; N′ = total number of classified patients.

The details of the cluster instance, as shown in Figure 4, was tested and classified as positive/negative. Out of 281 instances, 138 (49.11%) were classified as the 0 cluster instance, among them 47 (16.72%) were tested as negative, and 91 (32.38%) were tested as positive. One hundred and forty-three (50.88%) were classified as a cluster 1 instance from which 40 (14.23%) were tested as negative and 103 (36.65%) were tested as positive. In the final assessment, 51% were classified as positive and 49% instances as negative. The values of these classifications were used as input to the regression prediction phase.

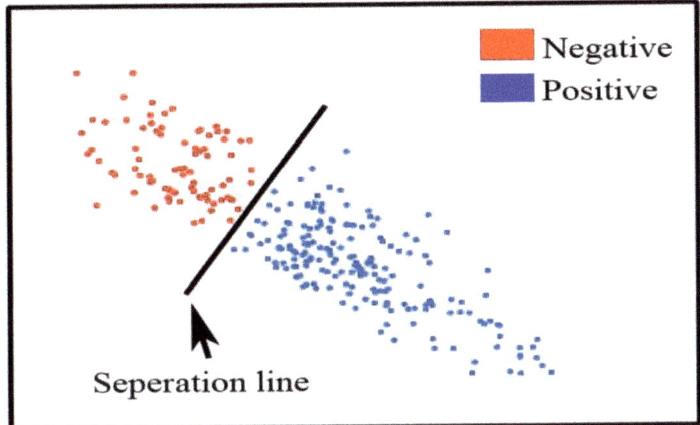

Figure 4. Evaluation of the Kmean clusters tested as positive and negative.

3.2. Rule Forecast Assessment

The predictive analysis represents the assessment for decision-making by determining the ratio of patient characteristics. The forecast analysis obtained in the study is graphically displayed in Figure 5a–g, and the 23 rules achieved through the rule classification measurements are described in Table 2 in terms of the patients' initial screening stage of healthcare.

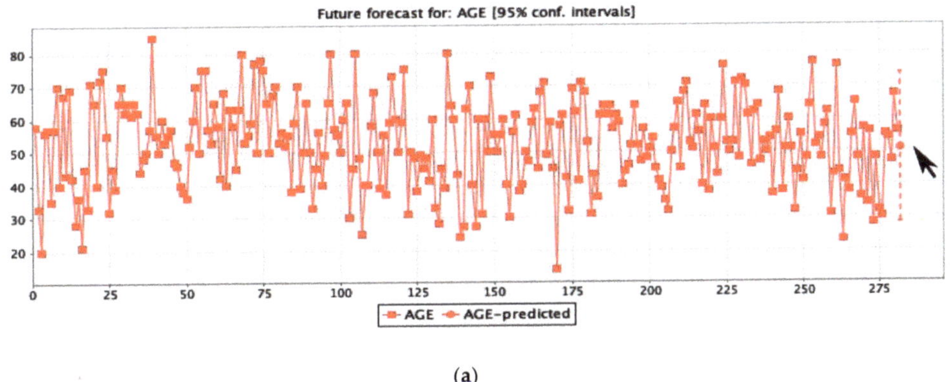

(a)

Figure 5. *Cont.*

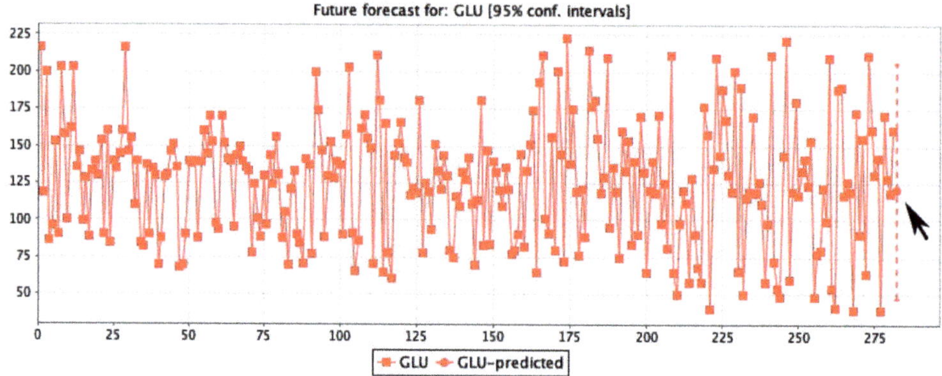

(**b**)

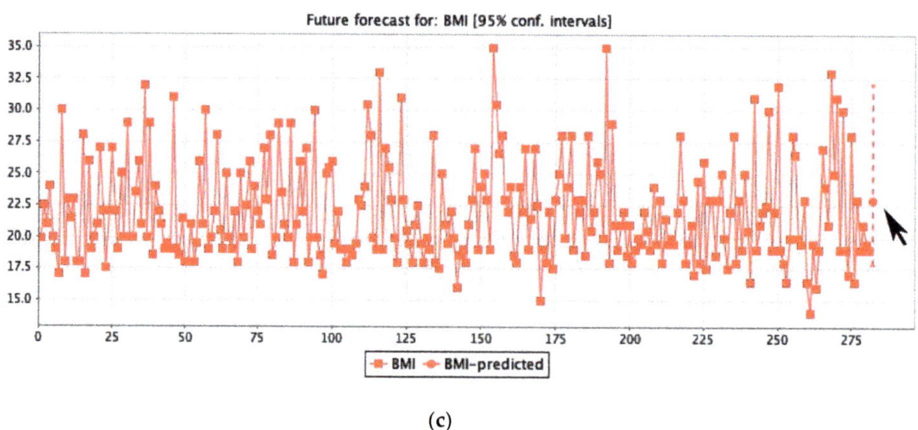

(**c**)

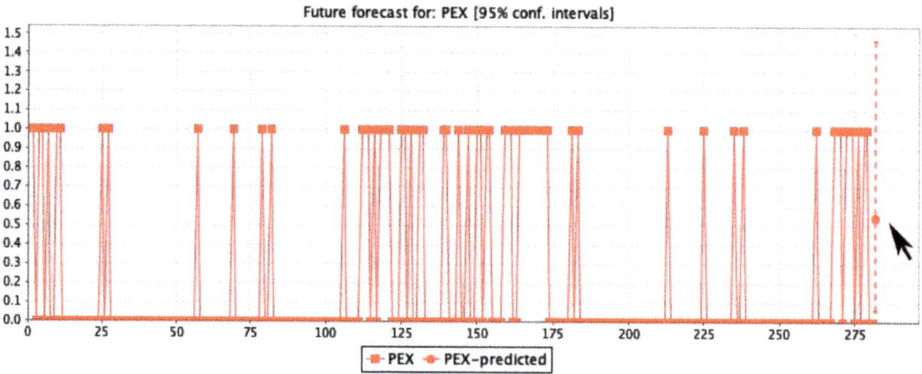

(**d**)

Figure 5. *Cont.*

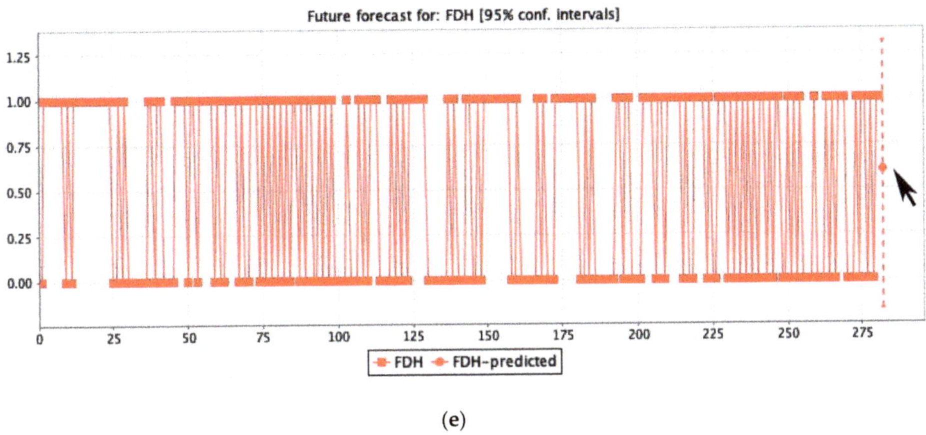

(**e**)

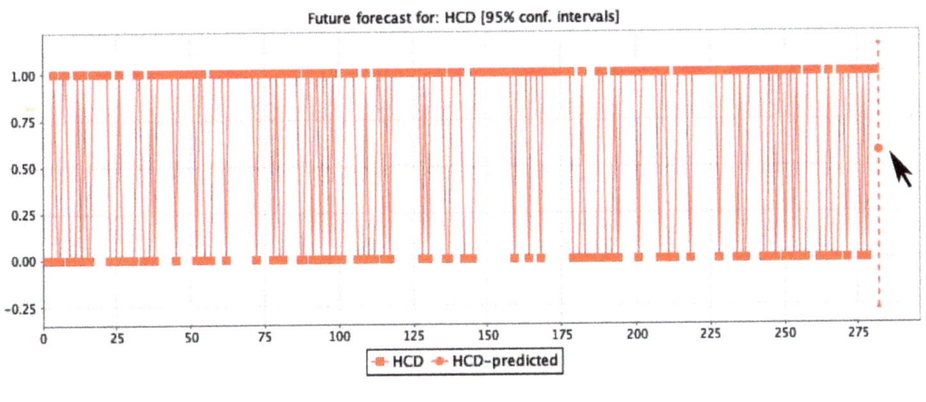

(**f**)

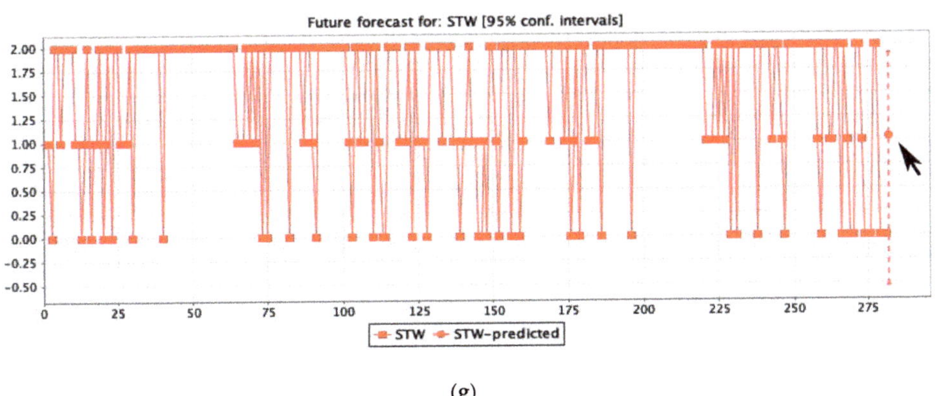

(**g**)

Figure 5. (a–g) The regression prediction assessment of the seven main features used for the analysis of clinical significance.

Table 2. Twenty-three if-then rules achieved from the classification analysis.

	Twenty-Three If-Then Rules Extracted from the Assessment Are:
Rule 1:	IF the patient's glucose level is (>101); THEN the patient is classified as tested positive with diabetes.
Rule 2:	IF the patient's glucose level is (>72); THEN the patient is classified as tested positive for diabetes, but the patient has to screen through the second stage test.
Rule 3:	IF the patient's blood pressure is (≤100); THEN the patient is classified as tested negative for diabetes but this case also depends on the glucose level of the patient, which takes patients for screening of the second stage.
Rule 4:	IF the patient's blood pressure is (<100); THEN the patient is classified as tested negative for diabetes but the patient has to go through the second stage of screening.
Rule 5:	IF the patient's (age ≤ 49) and (BMI ≤ 25) and the patient also has no diabetes in their family history; THEN the patient is classified as tested negative for diabetes.
Rule 6:	IF the patient's (age ≤ 34) and (BMI > 25) and the patient also has no diabetes in their family history and patient's diet is unbalanced; THEN the patient is classified as tested negative for diabetes.
Rule 7:	IF the patient's age is from (35 ≤ 49) and (BMI > 25) and the patient also has no diabetes in their family history and the patient's diet is unbalanced and the patient is without physical exercise; THEN the patient is classified as tested positive for diabetes.
Rule 8:	IF the patient's age is from (35 ≤ 49) and (BMI > 25), and the patient also has no diabetes in their family history, the patient's diet is unbalanced, and the patient is with physical exercise but has no history of cardiovascular disease; THEN the patient is classified as tested negative for diabetes.
Rule 9:	IF the patient's age is from (35 ≤ 49) and (BMI > 25), and the patient also has no diabetes in their family history, the patient's diet is unbalanced, and the patient is with physical exercise but has no history of cardiovascular disease; THEN the patient is classified as tested positive for diabetes.
Rule10:	IF the patient's age is (≤49) and (BMI > 25), and the patient also has no diabetes in their family history and the patient's diet is balanced; THEN the patient is classified as tested negative for diabetes.
Rule11:	IF the patient's age is (≤49) and (BMI ≤ 25), and the patient also has diabetes in their family history; THEN the patient is classified as tested negative for diabetes.
Rule12:	IF the patient's age is (≤49) and (BMI > 25), and the patient also has diabetes in their family history; THEN the patient is classified as tested positive for diabetes.
Rule13:	IF the patient's age is (>49) and (BMI ≤ 25), and the patient also has a high work stress but no diabetes in their family history; THEN the patient is classified as tested negative for diabetes.
Rule14:	IF the patient's age is (>49) and (BMI > 25), and the patient also has a high work stress but no diabetes in their family history; THEN the patient is classified as tested positive for diabetes.
Rule15:	IF the patient's age is (>49) and the patient has a high work stress, and also has diabetes in their family history; THEN the patient is classified as tested positive for diabetes.
Rule16:	IF the patient's age is (>49) and (BMI >25), and the patient's work stress is low and also has no diabetes in their family history but their diet is unbalanced; THEN the patient is classified as tested positive for diabetes.
Rule17:	IF the patient's age is (>49) and (BMI > 25), and the patient has no diabetes in their family history and has a balanced diet; THEN the patient is classified as tested negative for diabetes.
Rule18:	IF the patient's age is (>49) and (BMI > 25), and the patient's work stress is low but they have diabetes in their family history; THEN the patient is classified as tested positive for diabetes.

Table 2. *Cont.*

	Twenty-Three If-Then Rules Extracted from the Assessment Are:
Rule19:	IF the patient's age is (>49) and (BMI ≤ 25), and the patient has a low or medium work stress with hypertension and also their food is not balanced; THEN the patient is classified as tested positive for diabetes.
Rule20:	IF the patient is male with age (>49) and (BMI ≤ 25), and the patient has a low or medium work stress without hypertension and also their food is not balanced but they have diabetes in their family history with cardiovascular disease; THEN the patient is classified as tested positive for diabetes.
Rule21:	IF the patient is male with age (>49) and (BMI ≤ 25), and the patient has a low or medium work stress without hypertension and their diet is not balanced, and they have cardiovascular disease history in their family; THEN the patient is classified as tested negative for diabetes.
Rule22:	IF the patient is female with age (>49) and (BMI ≤ 25), and the patient has a low or medium work stress without hypertension and their diet is not balanced; THEN the patient is classified as tested negative for diabetes.
Rule23:	IF the patient's age is (>49) and (BMI ≤ 25), and the patient has a low or medium work stress with balanced diet; THEN the patient is classified as tested negative for diabetes.

The prediction assessment by logistic regression used in this study for clinical significance was analyzed by the confidence interval of 0.95%. The patient features used were age, blood glucose, body mass index, physical exercise, family history of diabetes, family cardiovascular history, and work stress by the M5 method in regression. The results of the forecast prediction for diabetes mellitus patients on the age feature show that patients up to 51 years could have a high death risk if the ratio of other features include a glucose level of 120.45 mmol/L, BMI ≥ 23, physical exercise between 0.5 to 0.6, family diabetes history of 0.6, cardiovascular stroke history of 0.61, and a work-stress ratio count of 1.08.

4. Discussion

In this study, a machine-learning technique was instigated on a data-mining platform with a dataset of 281 patients suffering from diabetes. The data was collected only from Nigeria for the assessment of diabetes mellitus prevalence by determining two rule classifiers (PART and Decision tables) on 10 non-invasive and easily accessible medical attributes/variables. They include age (age of the patient), gender (male and female), glucose level of the patient, body mass index of the patient, hypertension, history of cardiovascular disease, family history of diabetes, physical exercise, stress of work, and diet of the patient (healthy and unhealthy) to accurately measure diabetes mellitus ratio for rapid and precise screening of patients suffering with diabetes mellitus status along with other chronic disease symptoms.

Initially, during the assessment on the data mining platform (Weka), the dataset was divided into two parts for training and testing in a 20:80 percent ratio. Twenty percent of the training data was used to train the machine and assess the outcome. Whereas, 80 percent of the data was used for testing. Furthermore, a complete dataset of 281 patients was analyzed on the experimental mode of Weka for the final assessment of both classifiers together. The results of the Rule classification show the mean accuracy of 98.75% with an error rate of 0.02%. In addition, the mean kappa stats were 0.97%, true positive rate remained 0.97%, false positive rate 0.01%, precision 0.98%, recall 0.98%, F-matrix 0.98%, MCC 0.97%, ROC area ratio 0.99%, and PRC area ratio 0.99%.

The outcomes of the non-invasive medical features used in this study indicate this assessment can successfully help to predict the patients of diabetes and pre-diabetes without the need for preliminary laboratory tests. In addition, the 23 rules generated during the assessment clearly show the main features of individuals with diabetes. Therefore, this study raises the prediction that age is the underlying and root variable, followed by a family history of diabetes, body mass index, gender, work stress, physical exercise, diet lifestyle, hypertension, and cardiovascular family history. These implementations are useful for substantial epidemiological threats and low socioeconomic status regions around the world, such as Africa and other developing states.

The key strength of this study is its use of a unique approach to both classifiers with logistic regression assessment to identify and forecast diabetes mellitus prevalence. Moreover, the use of realistic health records collected from the four principal hospitals in the developing country of Nigeria where the prevalence proportion of diabetes in men and women is high and explicitly mentioned in the literature study. Hence, patients with diabetes mellitus can be screened by 23 generated rules. Diabetes mellitus can be controlled through organizing appropriate educational programs in developing countries to govern the widespread growth of diabetes mellitus. This can help people reduce the burden of health hitches through awareness-raising activities. The classification assessment proposed in this paper was set to test other well-known machine learning algorithms by the same data to evaluate and compare classification accuracy results. Table 3 and Figure 6 clearly show that PART and Decision table rule classifiers have been successful in clinically meaningful research.

Table 3. The rule classification average precision is compared to other machine learning classifiers based on the same dataset.

Method	Accuracy%	Mean%
PART rule	99.28	
Decision table rule	98.22	98.75%
MLP	73.82	
Discrim	77.54	
Logdisc	78.22	
KNN	94.29	
Logistic	85.35	
BayesNet	74.76	
NaïveBayes	76.35	
Random Forest	76.66	
LogitBoost	93.93	
J48	98.17	
SGD	76.62	
SMO	77.26	
ANN	89.84	
RBF	75.71	
FCM	94.78 [1]	

[1] It comprehensively compares the proposed classification results with the other machine learning classifiers on the same dataset.

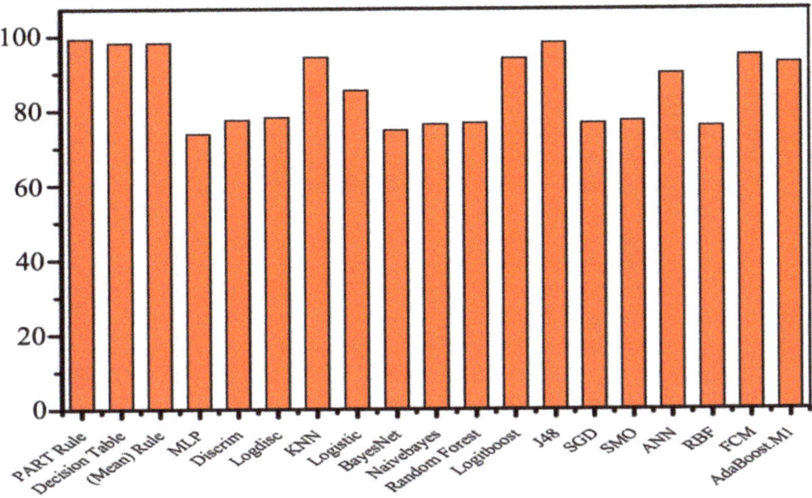

Figure 6. Comparisons of the rule classifier with other machine learning classifiers.

Limitation

The dataset was divided into a 20:80 percentages. Initially, 20% was utilized for machine training and 80% were used for testing. Furthermore, 100% with 281 instances were used in performing experiments on Weka to achieve the mean accuracy of both classifiers. This assessment study holds two limitations. One comprises a division of training and testing data for the meta-dataset and the second is the time taken to test the metadata for classification. If the metadata is analyzed on the same platform, the processing time can increase. However, it depends on the type of dataset used, the number of seeds input, and the number of experiments performed for acquiring the desired results.

5. Conclusions

This study implements the machine learning rule classifiers (PART and Decision table) on a data mining platform to identify possible diabetes and pre-diabetes in the initial clinical screening of a patient through logistic regression forecast assessment analysis. Two hundred and eighty-one diabetes mellitus patients have been analyzed with 10 easily available non-invasive medical features collected from four main hospitals located in northwestern Nigeria. The classification assessment accuracy was 98.75% and it was achieved through a set of 23-decision screening rules that can successfully influence accurate initial clinical screening of diabetes mellitus and pre-diabetes patients.

Additionally, the obtained Rules classified the most considerable risks and suggest that diabetes prevention and education programs can be applied in targeted community interventions. The study helps in the initial diagnosis of diabetes and reduces healthcare organization problems. Therefore, such a study is found extremely significant for the states and regions with extreme epidemic risk ratios and low socioeconomic status across the globe.

Author Contributions: Writing—Original draft preparation, methodology, software, and formal analysis have been done by M.N.S.; conceptualization, validation, data curation, and visualization, has done by M.N.S., M.U.M.; supervision, resources, project administration, and funding acquisition has done by R.J.; investigation has done by M.N.S., S.T.C., J.A.; finally writing—review and editing has carefully done by M.N.S., S.T.C., J.A., A.J.V.

Funding: This research work has been supported by NSFC Natural Science Foundation of Hebei province under grant of No. 61572420, No. 61472341, and No. 61772449.

Acknowledgments: We express our appreciation to "Yanshan University, Qinhuangdao, China" for accompanying us in this research.

Conflicts of Interest: The authors declare no conflict of interest.

Nomenclature

PART	Projective adaptive response theory
F-measure	Frequency matrix
CI	Confidence interval
MCC	Matthews's correlation coefficient
DR	Decision rules
DM	Diabetes mellitus
T2DM	Type 2 diabetes mellitus
GLU	Glucose level
BMI	Body mass index
HYP	Hypertension
HCD	History of cardiovascular disease
FDH	Family diabetes history
PEX	Physical exercise
STW	Work stress
DIT	Diet
LR	Logistic regression

References

1. Khoury, N.; Attal, F.; Amirat, Y.; Oukhellou, L.; Mohammed, S.; Khoury, N. Data-Driven Based Approach to Aid Parkinson's Disease Diagnosis. *Sensors* **2019**, *19*, 242. [CrossRef]
2. Melmed, S.; Polonsky, K.S.; Larsen, P.R.; Kronenberg, H. *Williams Textbook of Endocrinology*, 13th ed.; Elsevier: Amsterdam, The Netherlands, 2016.
3. Whiting, D.R.; Guariguata, L.; Weil, C.; Shaw, J. IDF Diabetes Atlas: Global estimates of the prevalence of diabetes for 2011 and 2030. *Diabetes Res. Clin. Pract.* **2011**, *94*, 311–321. [CrossRef]
4. Cho, N.H.; Shaw, J.E.; Karuranga, S.; Huang, Y.; da Rocha Fernandes, J.D.; Ohlrogge, A.W.; Malanda, B. IDF Diabetes Atlas: Global estimates of diabetes prevalence for 2017 and projections for 2045. *Diabetes Res. Clin. Pract.* **2018**, *138*, 271–281. [CrossRef] [PubMed]
5. Shamshirgaran, S.M.; Jorm, L.; Lujic, S.; Bambrick, H. Health related outcomes among people with type 2 diabetes by country of birth: Result from the 45 and Up Study. *Prim. Care Diabetes* **2019**, *13*, 71–81. [CrossRef] [PubMed]
6. Gan, D.; King, H.; Lefèbvre, P.; Mbanya, J.-C.; Silink, M.; Siminerio, L. *Diabetes Atlas*, 2nd ed.; Gent: Brussels, Belgium, 2015.
7. American Diabetes Association. 1. Improving Care and Promoting Health in Populations: Standards of Medical Care in Diabetes-2019. *Diabetes Care* **2019**, *42*, S7–S12. [CrossRef]
8. Tsobgny-Tsague, N.-F.; Lontchi-Yimagou, E.; Nana, A.R.N.; Tankeu, A.T.; Katte, J.C.; Dehayem, M.Y.; Bengondo, C.M.; Sobngwi, E. Effects of nonsurgical periodontal treatment on glycated haemoglobin on type 2 diabetes patients (PARODIA 1 study): A randomized controlled trial in a sub-Saharan Africa population. *BMC Oral Health* **2018**, *18*, 28. [CrossRef]
9. Lélis, V.-M.; Guzmán, E.; Belmonte, M.-V. A Statistical Classifier to Support Diagnose Meningitis in Less Developed Areas of Brazil. *J. Med. Syst.* **2017**, *41*, 145. [CrossRef] [PubMed]
10. Awad, S.F.; O'Flaherty, M.; Critchley, J.; Abu-Raddad, L.J. Forecasting the burden of type 2 diabetes mellitus in Qatar to 2050: A novel modeling approach. *Diabetes Res. Clin. Pract.* **2018**, *137*, 100–108. [CrossRef]
11. Choi, S.B.; Kim, W.J.; Yoo, T.K.; Park, J.S.; Chung, J.W.; Lee, Y.; Kang, E.S.; Kim, D.W. Screening for prediabetes using machine learning models. *Comput. Math. Methods Med.* **2014**, *2014*, 618976. [CrossRef] [PubMed]
12. Talaei-Khoei, A.; Wilson, J.M.; Kazemi, S.-F. Period of Measurement in Time-Series Predictions of Disease Counts from 2007 to 2017 in Northern Nevada: Analytics Experiment. *JMIR Public Heal Surveill.* **2019**, *5*, e11357. [CrossRef]

13. Olivera, A.R.; Roesler, V.; Iochpe, C.; Schmidt, M.I.; Vigo, Á.; Barreto, S.M.; Duncan, B.B. Comparison of machine-learning algorithms to build a predictive model for detecting undiagnosed diabetes—ELSA-Brasil: Accuracy study. *Sao Paulo Med. J.* **2017**, *135*, 234–246. [CrossRef] [PubMed]

14. Sohail, M.N.; Jiadong, R.; Uba, M.M.; Irshad, M. *A Comprehensive Looks at Data Mining Techniques Contributing to Medical Data Growth: A Survey of Researcher Reviews*; Springer: Singapore, 2019; pp. 21–26.

15. Kaur, P.; Sharma, N.; Singh, A.; Gill, B. CI-DPF: A Cloud IoT based Framework for Diabetes Prediction. In Proceedings of the 2018 IEEE Annual Information Technology, Electronics and Mobile Communication Conference, Columbia, Canada, 3 November 2018; pp. 654–660.

16. Mahmud, S.M.H.; Hossin, M.A.; Ahmed, M.R.; Noori, S.R.H.; Sarkar, M.N.I. Machine Learning Based Unified Framework for Diabetes Prediction. In Proceedings of the 2018 International Conference on Big Data Engineering and Technology (BDET 2018), Chengdu, China, 27 August 2018; pp. 46–50.

17. Srikanth, P.P.; Nilofer, V.; Siddiqui, I.; Dasari, P.; Ambica, B.; Venkata, V.B.V.E. Characteristic evaluation of diabetes data using clustering techniques. *Int. J. Comput. Sci. Netw. Secur.* **2008**, *8*, 244–251.

18. Okpor, M.D. Prognostic Diagnosis of Gestational Diabetes Utilizing Fuzzy Classifier. *Int. J. Comput Sci. Netw. Secur.* **2015**, *15*, 44.

19. Humayun, A.; Niaz, M.; Umar, M.; Mujahid, M. Impact on the Usage of Wireless Sensor Networks in Healthcare Sector. *Int. J. Comput. Sci. Netw. Secur.* **2017**, *17*, 102–105.

20. Atmini, D.; Dwi, L.; Eminugroho, R. Sensitivity Analysis of Goal Programming Model for Dietary Menu of Diabetes Mellitus Patients. *Int. J. Model. Optim.* **2017**, *7*, 7–14. [CrossRef]

21. Raimundo, M.S.; Okamoto, J., Jr. Application of Hurst Exponent (H) and the R/S Analysis in the Classification of FOREX Securities. *Int. J. Model. Optim.* **2018**, *8*, 116–124. [CrossRef]

22. Alotaibi, M. Investigating the Role of Social Robot in improving diabetic Children Management and awareness. *Int. J. Comput. Sci. Netw. Secur.* **2017**, *17*, 121.

23. Khawaja, A.P.; Cooke Bailey, J.N.; Wareham, N.J.; Scott, R.A.; Simcoe, M.; Igo, R.P.; Song, Y.E.; Wojciechowski, R.; Cheng, C.-Y.; Khaw, P.T.; et al. Genome-wide analyses identify 68 new loci associated with intraocular pressure and improve risk prediction for primary open-angle glaucoma. *Nat. Genet.* **2018**, *50*, 778–782. [CrossRef]

24. Dian, S.; Rahmadi, R.; van Laarhoven, A.; Ganiem, A.R.; van Crevel, R. Predicting Mortality of Tuberculous Meningitis. *Clin. Infect. Dis.* **2018**, *67*, 1954–1955. [CrossRef] [PubMed]

25. Samy, G.; Gamal, N. A Framework for Social Network-Based Dynamic Modeling and Prediction of Communicable Disease. *Int. J. Model. Optim.* **2019**, *9*, 30–33. [CrossRef]

26. Lamwong, J.; Pongsumpun, P. Age Structural Model of Zika Virus. *Int. J. Model. Optim.* **2018**, *8*, 17–23. [CrossRef]

27. Verdial, F.; Madtes, D.; Hwang, B.; Mulligan, M.; Odem-Davis, K.; Waworuntu, R.; Wood, D.; Farjah, F. A Prediction Model for Nodal Disease among Patients with Non-Small Cell Lung Cancer. *Ann. Thorac. Surg.* **2019**. [CrossRef] [PubMed]

28. Kasiakogias, A.; Tsioufis, C.; Dimitriadis, K.; Konstantinidis, D.; Koutra, E.; Kyriazopoulos, K.; Kyriazopoulos, I.; Liatakis, I.; Mantzouranis, M.; Philippou, C.; et al. P1540Comparison of the European Society of Hypertension stratification and European Society of Cardiology HeartScore for prediction of coronary artery disease and stroke in essential hypertension. *Eur. Heart J.* **2018**, *39*, 1540. [CrossRef]

29. Mustaqeem, A.; Anwar, S.M.; Majid, M.; Khan, A.R. Wrapper method for feature selection to classify cardiac arrhythmia. In Proceedings of the 39th Annual International Conference of the IEEE Engineering in Medicine and Biology Society, Jeju Island, Korea, 11–15 July 2017; pp. 3656–3659.

30. Romanowski, K.; Balshaw, R.F.; Benedetti, A.; Campbell, J.R.; Menzies, D.; Ahmad Khan, F.; Johnston, J.C. Predicting tuberculosis relapse in patients treated with the standard 6-month regimen: An individual patient data meta-analysis. *Thorax* **2019**, *74*, 291–297. [CrossRef] [PubMed]

31. Alfian, G.; Syafrudin, M.; Ijaz, M.; Syaekhoni, M.; Fitriyani, N.; Rhee, J. A Personalized Healthcare Monitoring System for Diabetic Patients by Utilizing BLE-Based Sensors and Real-Time Data Processing. *Sensors* **2018**, *18*, 2183. [CrossRef] [PubMed]

32. Ijaz, M.; Alfian, G.; Syafrudin, M.; Rhee, J. Hybrid Prediction Model for Type 2 Diabetes and Hypertension Using DBSCAN-Based Outlier Detection, Synthetic Minority Over Sampling Technique (SMOTE), and Random Forest. *Appl. Sci.* **2018**, *8*, 1325. [CrossRef]

33. Kieviet, A. Werkzeuge der digitalen Transformation. In *Lean Digital Transformation*; Springer: Berlin/Heidelberg, Germany, 2019; pp. 57–159.

34. Witten. Weka—Data Mining with Open Source Machine Learning Software in Java. Weka. 2016. Available online: https://www.cs.waikato.ac.nz/ml/weka/ (accessed on 23 May 2018).

35. Fallah, M.; Niakan Kalhori, S.R. Systematic Review of Data Mining Applications in Patient-Centered Mobile-Based Information Systems. *Healthc. Inform. Res.* **2017**, *23*, 262. [CrossRef] [PubMed]

36. Padillo, F.; Luna, J.M.; Ventura, S. A Grammar-Guided Genetic Programing Algorithm for Associative Classification in Big Data. *Cognit. Comput.* **2019**, 1–16. [CrossRef]

37. González, J.; Ortega, J.; Damas, M.; Martín-Smith, P.; Gan, J.Q. A new multi-objective wrapper method for feature selection—Accuracy and stability analysis for BCI. *Neurocomputing* **2019**, *333*, 407–418. [CrossRef]

38. Muhammad, M.U.; Asiribo, O.E.; Noman, S.M. Application of Logistic Regression Modeling Using Fractional Polynomials of Grouped Continuous Covariates. *Niger. Stat. Soc.* **2017**, *1*, 144–147.

39. Dubey, R.; Makwana, R.R.S. Computer-Assisted Valuation of Descriptive Answers Using Weka with RandomForest Classification. In *Proceeding of the Second International Conference on Microelectronics, Computing & Communication Systems (MCCS 2017)*, 76th ed.; LNEE, Ed.; Springer: Singapore, 2019; pp. 359–366.

40. Rani, R.U.; Kakarla, J. Efficient Classification Technique on Healthcare Data. In *Progress in Advanced Computing and Intelligent Engineering*, 713rd ed.; AISC, Ed.; Springer: Singapore, 2019; pp. 293–300.

Article

A Machine Learning-based Pipeline for the Classification of CTX-M in Metagenomics Samples

Diego Ceballos [1,2,*], Diana López-Álvarez [3], Gustavo Isaza [2,*], Reinel Tabares-Soto [1], Simón Orozco-Arias [1,2] and Carlos D. Ferrin [4]

[1] Department of Electronics and Automatization, Universidad Autónoma de Manizales, Manizales 1700, Colombia; rtabares@autonoma.edu.co (R.T.-S.); simon.orozco.arias@gmail.com (S.O.-A.)

[2] Universidad de Caldas, Manizales 1700, Colombia

[3] Universidad Nacional de Colombia-Palmira, Palmira 763531, Colombia; dianalopez430@gmail.com

[4] Universidad del Valle, Cali 760001, Colombia; cdfbdex@gmail.com

* Correspondence: diegoh.ceballosl@autonoma.edu.co (D.C.); gustavo.isaza@ucaldas.edu.co (G.I.)

Received: 16 February 2019; Accepted: 8 April 2019; Published: 24 April 2019

Abstract: Bacterial infections are a major global concern, since they can lead to public health problems. To address this issue, bioinformatics contributes extensively with the analysis and interpretation of in silico data by enabling to genetically characterize different individuals/strains, such as in bacteria. However, the growing volume of metagenomic data requires new infrastructure, technologies, and methodologies that support the analysis and prediction of this information from a clinical point of view, as intended in this work. On the other hand, distributed computational environments allow the management of these large volumes of data, due to significant advances in processing architectures, such as multicore CPU (Central Process Unit) and GPGPU (General Propose Graphics Process Unit). For this purpose, we developed a bioinformatics workflow based on filtered metagenomic data with Duk tool. Data formatting was done through Emboss software and a prototype of a workflow. A pipeline was also designed and implemented in bash script based on machine learning. Further, Python 3 programming language was used to normalize the training data of the artificial neural network, which was implemented in the TensorFlow framework, and its behavior was visualized in TensorBoard. Finally, the values from the initial bioinformatics process and the data generated during the parameterization and optimization of the Artificial Neural Network are presented and validated based on the most optimal result for the identification of the CTX-M gene group.

Keywords: machine learning; metagenomics; bioinformatics; CTX-M

1. Introduction

Within the field of bioinformatics, researchers use metagenomics approaches to characterize microbial genomes directly isolated from the environment [1]. For this, new sequencing technologies generate large volumes of data to be analyzed, due to the abundant varieties of species that can be found in metagenomics samples, which are characterized by sequences of short length and high complexity. In addition, with the possibility of discovering new species, the problem of taxonomic assignment of reads of short DNA sequences becomes extremely challenging [2]. In this respect, metagenomics is considered as the field of study of many genomes in different environments that may even be compartments or regions of living beings, such as mucous membranes and intestines, among others. Therefore, metagenomics is a challenge for computer science researchers who seek to develop methods to understand such amount of genetic information [3]. Concerning the area of computational intelligence, this work deals with a technique already known and validated with artificial neural networks. According to [3] Soueidan and Hayssam (2016), machine learning techniques currently offer a large set of promising tools to build predictive models for the classification of biological data. These tools are built under different frameworks offering the possibility of implementing supervised and unsupervised techniques (clustering), among others.

CTX-M-type enzymes are a group of class A extended-spectrum β-lactamases (ESBLs) that are rapidly spreading among Enterobacteriaceae worldwide. The first recognition of the appearance of CTX-M β-lactamases occurred almost simultaneously in Europe and South America in early 1989. The first publication to recognize an ESBL from the CTX-M group was a report presenting a species of *E. coli* resistant to cefotaxime but susceptible to ceftazidime, isolated from the ear of a four-month-old child suffering from otitis media in Munich [4].

At the regional level, the Manizales Antibiotic Resistance Group (GRAM) is in charge of presenting the accumulated antibiotic resistance data of the main hospitals in the city. Among total isolates from patients in intensive care units, non-intensive care units and emergencies, the main bacteria identified are Enterobacteriaceae such as *Escherichia coli*, *Klebsiella pneumoniae*, and *Eneterobacter cloacae*, among others. All of these species display the capacity to carry ESBL genes of the CTX-M group. In addition, according to the antibiotic susceptibility analyses carried out by different clinics in the city, resistance to cefotaxime (cephalosporin with a broad hydrolysable spectrum by CTX-M) ranges between 15% and 35% [5]. This means that, in Manizales, up to one out of every three isolates of this bacterial group is suspected of carrying a CTX-M-type ESBL. The high frequency of this type of ESBL in our context highlights the importance of this type of developments for antibiotic surveillance processes based on metagenomic data.

The validation of this pipeline allows us to extend this analysis for other important genes such as *TEM, SHV, metalloenzymes, carbapenemases* that are probably prevalent in our regional context, considering the characteristics of the population, the clinical management protocols of patients and health, and asepsis in operating rooms. Since this is a common problem, the development of a pipeline that allows the identification of resistance variants becomes a fundamental step in the establishment of a modern antibiotic surveillance system. The subsequent goal of this study will be to test this development on metagenomic data derived from the surveillance process, in collaboration with research groups in this field.

1.1. Metagenomics

According to the National Center for Biotechnology Information (NCBI) [6], metagenomics is an area of bioinformatics that has evolved significantly in the last ten years, contributing on a large scale to microbiology. In the same manner, this relatively new "omic" science has made surprising discoveries in microbial taxonomy, revealing new capabilities and functionalities of different biomes [7].

Metagenomics is analyzed through computation and bioinformatics, especially with the use of different information discovery techniques. From this field, we try to discover patterns within this data to extract information that may be relevant for biologists, pharmacologists, chemists and/or bioinformaticians. This information contributes to the solution of different pathologies related to microbial attacks.

New techniques have been developed to analyze large volumes of information from large amounts of metagenomic data, being big data and machine learning the most widely used [8]. These techniques use distributed computational environments of large capacity that allow more efficient processing and reduce computing times in a significant way.

1.2. Machine Learning

Machine learning seeks to answer a very concrete question: How can we build computer systems that automatically improve with experience, and what fundamental laws govern this teaching process? [9]

Through this discipline, it is possible to implement new methods that help researchers in making new findings. Machine learning techniques are used, for example, to learn about models of gene expression in cells and other applications in bioinformatics, more specifically in metagenomics [10]. One can talk about three types of algorithms within the current machine learning techniques:

Supervised: Data training consists of labeled entries and known outputs that the machine analyzes while relabeling. There are many applications of supervised algorithms in bioinformatics to solve problems [11], which are based on information from adequately characterized genes.

Unsupervised: This type of analysis of unlabeled and categorized data is based on similarities that have been identified. In this case, the machine can cluster the data based on shared characteristics. Techniques that use unsupervised algorithms are often used for problems in which humans cannot clearly infer patterns, that is, it requires exhaustive observation to identify such patterns. It is also a technique that allows determining behaviors based on different interpretations.

Semi-supervised: This analysis refers to a combination of the two previously mentioned techniques. It is used in large data sizes when the labels of some of these data are known. Unsupervised learning is based on the analysis of unlabeled data to group them, while techniques of supervised learning are used to predict the labels of this group formed by the first technique. Artificial Neural Networks (ANN) are a known approach to address complex problems, as neural networks can be implemented at the hardware or software level and, in turn, can use a variety of topologies and learning algorithms.

2. Materials and Methods

2.1. Selection of the CTX-M and Metagenome Baseline Reference Database for the Study

First, we based our selection on previous work by [12] Núñez in 2016 (unpublished data), where all the CTX-M reported groups are already considered. After a review of the state of the art, we consolidated the CTX-M database, previously filtered by the analysis of phylogenetic trees carried out by [12] Núñez. Subsequently, the reference metagenome to be studied was selected through a search in the EBI-Metagenomics database (https://www.ebi.ac.uk/metagenomics/), considering the high probability that the *CTX-M* gene was present. We reviewed the following four metagenomes and selected only one as input to develop the prototype:

1. https://www.ebi.ac.uk/metagenomics/projects/ERP001506
2. https://www.ebi.ac.uk/metagenomics/projects/ERP020191
3. https://www.ebi.ac.uk/metagenomics/projects/ERP016968
4. https://www.ebi.ac.uk/metagenomics/projects/ERP009131

The metagenome selected was antibiotic resistance within the preterm infant gut (https://www.ebi. ac.uk/ena/data/view/PRJEB15257). Upon selection of the reference metagenome, we filtered the data by following the pipeline described in Figure 1. The filtered metagenomics data was then prepared and machine learning techniques were applied according to the computational pipeline shown in Figure 2, where we assessed the accuracy and cost of the artificial neural network. A brief description is as follows: the filtered metagenome from the first pipeline is provided as input; the data are transformed by the conversion of nucleotide to binaries and the resulting binarized data are input to the ANN (Artificial Neural Network); the ANN is implemented; and accuracy and cost metrics are assessed.

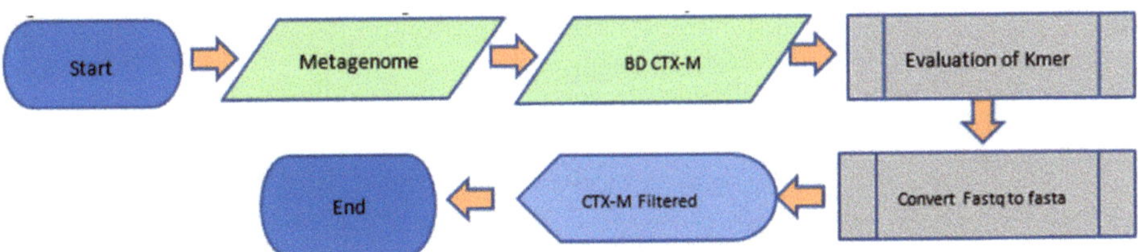

Figure 1. Details of the bioinformatic pipeline.

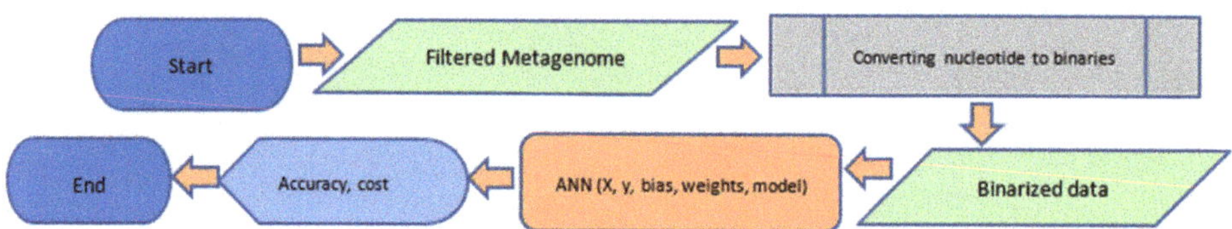

Figure 2. Details of the computational pipeline.

We mapped the CTX-M reference database to the sample metagenome using Duk tool (Li, Mingkun, et al., 2018) to eliminate information not relevant for the study. We obtained a consolidated CTX-M database with a total of 211 reference sequences in FASTA (file format for bioinformatics data). As initial mapping parameters, we used k-mers of 16 (default) and 63 for test mappings. Next, we optimized mapping parameters following Algorithm 1.

Algorithm 1. Bioinformatic pipeline for filtering and formatting input data.

Parameterize the initial mapping with Duk using odd K-mers.
Execute tests using different K-mers.
Name: Pre-filter CTX-M
Start
    For k-mer values between 17 and 65
    Do
      Execute duk with each k-mer against the reference database
      Save results in a single file "duk_results"
    Finish do
    Best_K-mer < 0
    Best p-value < 0
    For each line in "duk_results" file
    Do
      Find p-value of each k-mer
      If (P-value found is larger than Best p-value)
        Best p-value < p-value found
        Best_K-mer < k-mer found
      End if
      Convert output file of best k-mer to FASTA format
      Format the FASTA file for the ANN (X, y)
      For each end of CTX-M sequence
      Do
        Separate CTX-M group from each sequence.
    Finish do
End

Based on the initial analysis, k-mers 17, 19 and 21 were found to be the best. Additionally, we validated the results through an NCBI BLAST search of the contig obtained after adjusting the k-mer to 17 and 19 to conclusively verify that this sequence corresponds to bacteria with the CTX-M gene. The pipeline can be downloaded here:

https://github.com/dhcl1580/machinelearniginmetagenomicstesis.

2.2. Defining an Optimal Neural Network Architecture

An exhaustive review of the existing literature was performed to define the architecture of the neural network for metagenomics. We evaluated different machine learning models focused on improving the precision of the techniques applied in neural networks, such as random forest, or algorithms based on decision trees [13]. None of the studies reviewed take into account a particular architecture, whereby the main goal is to obtain a reduction in the cost function to guarantee that the neural network apprenticeship is being carried out. Conversely, this study proposes an architecture of a multi-layer perception neuronal network (Figure 3), because of the importance of the high sensitivity that different neurons show in each of their layers concerning the activation functions, weights, and epochs. This interaction allows considering more parameters when training and validating such an architecture, taking into account its performance [14].

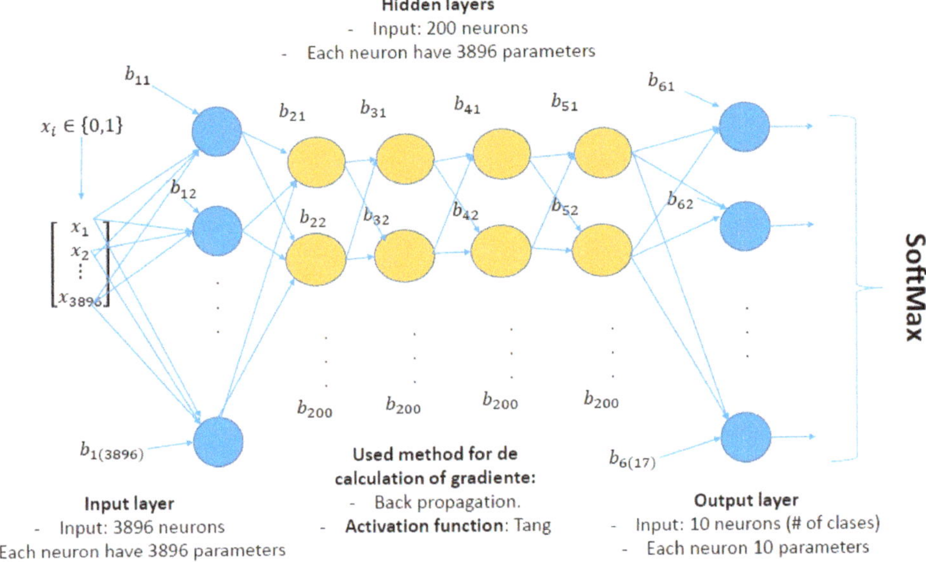

Figure 3. Details of the architecture of ANN (Artificial Neural Network).

2.3. Data Standardization for the Neural Network

To establish an appropriate training dataset for the proposed neuronal network, we developed a routine in Python 3 in charge of normalizing the data obtained, where basically a binarization of the CTX-M nucleotide sequences is carried out. All sequences are standardized to the value of the longest identified sequence, and additional spaces are defined by the value N. The result is the file "dataGen.csv", where a total of 3896 values are generated for X and the 10 groups of CTX-M (Table 1). The 10 most representative classes were selected to ensure a uniform distribution of classes for stratified cross validation in Stage 2 (validation). Initially, there were 17 classes from which only those with sequences represented at least four times within the test and validation dataset were selected. Each of the 10 classes corresponds to the following CTX-M groups, respectively (Table 1).

Table 1. CTX-M group and correspond class selected for the study.

Group CTX-M	Class
1.0	0
9.0	1
14.0	2
15.0	3
22.0	4
24.0	5
27.0	6
55.0	7
59.0	8
65.0	9

3. Analysis of Results

3.1. Analysis of the Graph Resulting from the ANN

Figure 4 shows how the graph of the ANN is built. In this graph, it is possible to observe how the nodes are distributed and how these interaction to the process data.

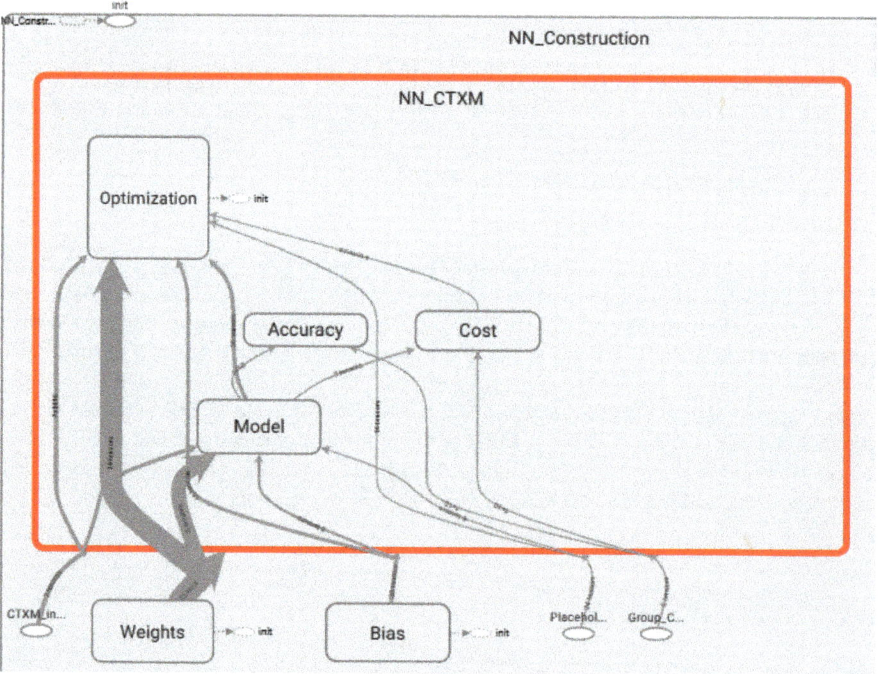

Figure 4. Details of the ANN (Artificial Neural Network) components and the cost, accuracy, optimization and model definition tensors.

3.2. Training Stage Over CPU an GPGPU

The activation functions tanh and sigmoid were experimented with RELU (Rectified Linear Units), where the parameters LEARNING_RATE, TRAINING_EPOCHS, and HIDDEN_SIZE were varied, obtaining the results presented below for each function. Table 2 shows the parameters that varied in each experiment. The Figures 5–7 show the correspond graphics.

Table 2. Summary of target values during the training stage under CPU (Central Process Unit).

Activation Function	LEARNING_ RATE	TRAINING_ EPOCH	HIDDEN_ SIZE	Initial Cost Value	Final Cost Value	Accuracy of Initial Training	Accuracy of Final Training	Precision Test
Tanh	0.001	400	200	2.17	0.80	0.260	0.960	0.879
Sigmoid	0.001	400	200	2.19	1.61	0.030	0.680	0.698
RELU	0.001	300	200	2.19	0.00	0.110	1	1

The best values were obtained using the tanh activation function in this experiment.

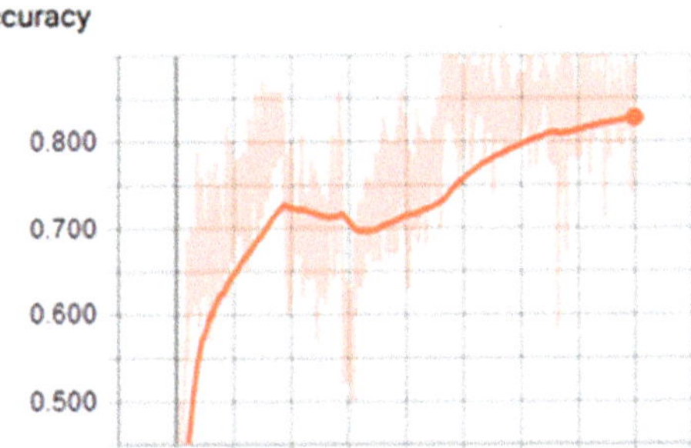

Figure 5. Values of accuracy using tanh function over CPU (Central Process Unit).

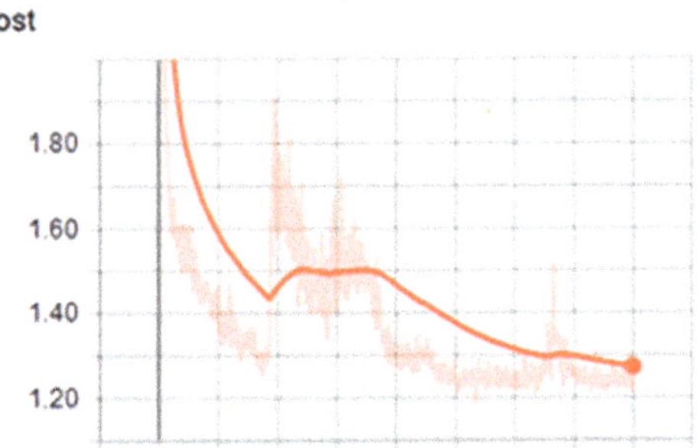

Figure 6. Values of cost using tanh function over CPU (Central Process Unit).

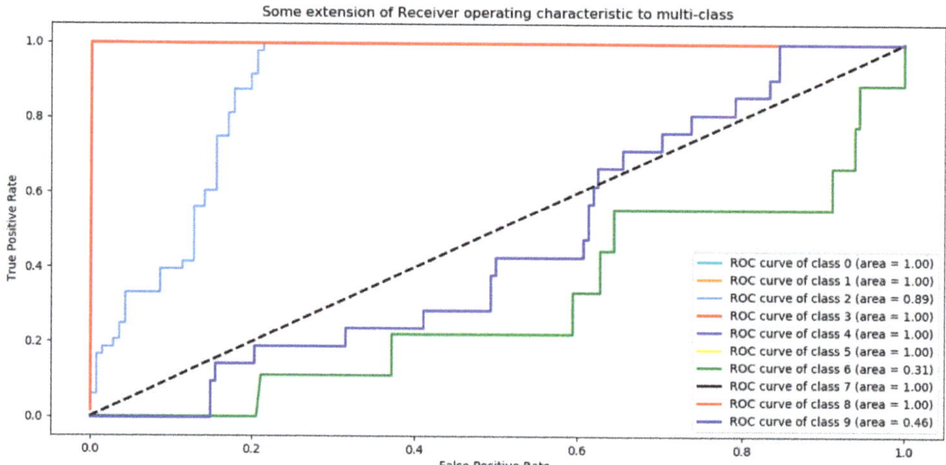

Figure 7. ROC (Receiver operating characteristics) analysis for the tanh activation function over CPU (Central Process Unit).

The best values were obtained using the tanh activation function in the other step, the Table 3 show the values ant the Figures 8–10 show the correspond graphics.

Table 3. Summary of target values during the training stage under GPU (Graphics Process Unit).

Activation Function	LEARNING_ RATE	TRAINING_ EPOCH	HIDDEN_ SIZE	Initial Cost Value	Final Cost Value	Accuracy of Initial Training	Accuracy of Final Training	Precision Test
Tanh	0.001	400	200	2.16	0.84	0.380	0.920	0.909
Sigmoid	0.001	400	200	2.20	1.67	0.440	0.560	0.628
RELU	0.001	300	200	1.90	1.00	0.590	1	1

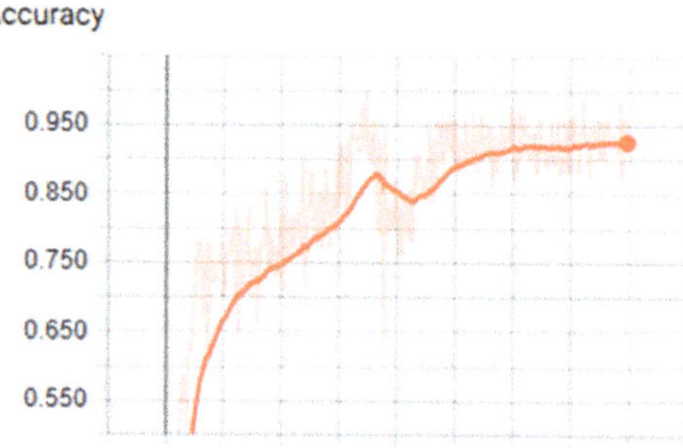

Figure 8. Values of accuracy using tanh function over GPU (Graphics Process Unit).

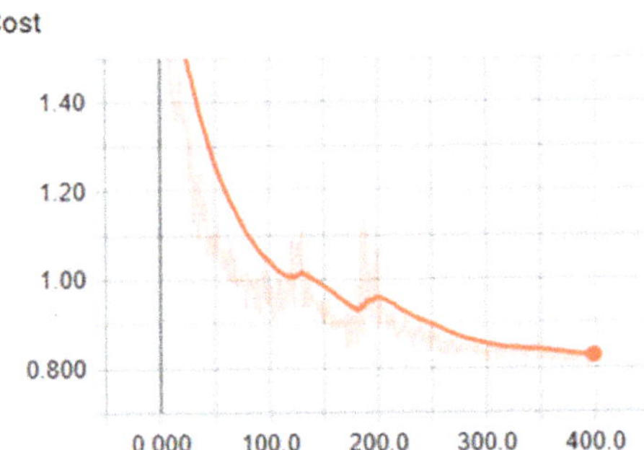

Figure 9. Values of cost using tanh function over GPU (Graphics Process Unit).

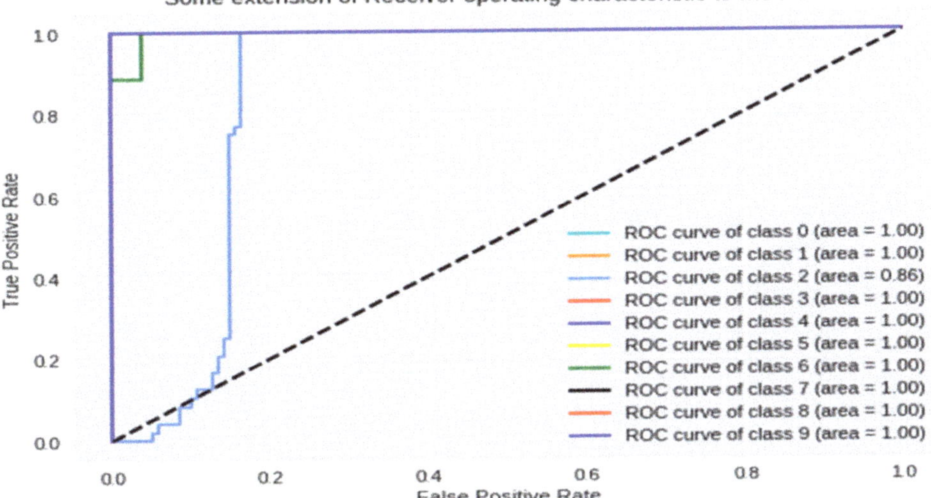

Figure 10. ROC (Receiver operating characteristics) analysis for the tanh activation function over GPU (Graphics Process Unit).

4. Discussion

4.1. Conclusions for the Tanh Activation Function

We found that the ANN showed the most optimal behavior under the tanh activation function for the training stage. The reference value was 0.879 for the precision test that varied the training epoch and hidden size parameters. Precision and cost behaviors were as expected, considering that the cost decreased and the precision increased for all the evaluations proposed under different parameters. Another relevant conclusion is that, according to the ROC analysis, the classes that are least likely to be identified under these ANN parameters are classes 2 and 6.

4.2. Conclusions About the Dataset

Regarding the dataset, we can conclude that, for future work, it is advisable to consider more CTX-M contigs. In this study, the 10 most representative groups were considered, yet some of the groups were not representative enough to be able to carry out a stratified cross validation. This was particularly true for the experimentation in the validation stage, in which 20% of the initial dataset was used for this validation. Regarding the dataset, we can conclude that more CTX-M contigs should be considered for future studies.

4.3. Perspective

In a future study, we propose to validate a more significant number of metagenomes corresponding to the geographical area of influence, aiming to support the design of public policies related to the prevention and detection of infectious diseases. To corroborate the final results more accurately, other types of metrics, especially histograms, would be considered, taking advantage of the fact that they can be generated by the TensorBoard tool. Finally, we recommended to continue with the training process with other genes such as *TEM, SHV, metalloenzymes, carbapenemases*, so that this software can identify a higher number of infectious diseases with the same characteristics.

Author Contributions: Conceptualization: D.C., D.L.A., G.I. and C.D.F.; formal analysis: D.L.A., G.I.E., S.O.A. and C.D.F.; investigation: D.C.; methodology: D.C.; software: S.O.A., R.T.S. and C.D.F.; supervision: D.L.A. and G.I.; writing–original draft: D.C.; writing—review and editing: D.L.A., G.I., S.O.A. and C.D.F.

Funding: This research was funded by Universidad Autónoma de Manizales.

Conflicts of Interest: The authors declare no conflict of interest.

References

1. Hoff, K.J.; Tech, M.; Lingner, T.; Daniel, R.; Morgenstern, B.; Meinicke, P. Gene prediction in metagenomic fragments: A large scale machine learning approach. *BMC Bioinform.* **2008**, *9*, 217. [CrossRef] [PubMed]
2. Rasheed, Z.; Rangwala, H. Metagenomic Taxonomic Classification Using Extreme Learning Machines. *J. Bioinform. Comput. Biol.* **2012**, *10*, 1250015. [CrossRef] [PubMed]
3. Soueidan, H.; Nikolski, M. Machine learning for metagenomics: Methods and tools. *arXiv* **2015**, arXiv:1510.06621. [CrossRef]
4. Cantón, R.; González-Alba, J.M.; Galán, J.C. CTX-M enzymes: origin and diffusion. *Front. Microbiol.* **2012**, *3*, 110. [CrossRef] [PubMed]
5. Salazar, J.D.; Loaiza, S.; Ibáñez, J.P.; Hernandez, J.S. Primera mirada a la resistencia antibiótica de la ciudad de Manizales. Segundo Simposio Regional de Resistencia Antibiótica–Eje Cafetero, 2018. Universidad de Manizales, noviembre 3 de 2018.
6. Thomas, T.; Gilbert, J.; Meyer, F. Metagenomics—A guide from sampling to data analysis. *Microb. Inform. Exp.* **2012**, *2*, 3. [CrossRef] [PubMed]
7. Johnson, J.; Jain, K.; Madamwar, D. 2—Functional Metagenomics: Exploring Nature's Gold Mine. In *Current Developments in Biotechnology and Bioengineering*; Elsevier: Amsterdam, The Netherlands, 2017; pp. 27–43. ISBN 9780444636676. Available online: http://www.sciencedirect.com/science/article/pii/B978044463667600002X (accessed on 11 October 2018).
8. Ma, C.; Zhang, H.H.; Wang, X. Machine learning for Big Data analytics in plants. *Trends Plant Sci.* **2014**, *19*, 798–808. [CrossRef] [PubMed]
9. Mitchell, T.M. *The Discipline of Machine Learning. CMU-ML-06-108*; School of Computer Science, Carnegie Mellon University: Pittsburgh, PA, USA, 2006.
10. Vervier, K.; Mahé, P.; Tournoud, M.; Veyrieras, J.B.; Vert, J.P. Large-scale Machine Learning for Metagenomics Sequence Classication. *Bioinformatics* **2015**, *32*, 1023–1032. [CrossRef] [PubMed]
11. Lu, P.; Abedi, V.; Mei, Y.; Hontecillas, R.; Philipson, C.; Hoops, S.; Carbo, A.; Bassaganya-Riera, J. *Emerging Trends in Computational Biology, Bioinformatics, and Systems Biology*; Elsevier: Amsterdam, The Netherlands, 2015; ISBN 9780128025086.
12. Nuñez, A. Anábioimutendifetide blaCTX-M. 2016.
13. Krachunov, M.; Sokolova, M.; Simeonova, V.; Nisheva, M.; Avdjieva, I.; Vassilev, D. Quality of Different Machine Learning Models In Error Discovery For Parallel Genome Sequencing. *Comptes Rendus De L Academie Bulgare Des Sciences* **2017**, *71*, 922–929.

14. Zeng, X.; Yeung, D.S. Sensitivity analysis of multilayer perceptron to input and weight perturbations. *IEEE Trans. Neural Netw.* **2001**, *12*, 1358–1366. [CrossRef] [PubMed]

Review

A Review of Computational Methods for Clustering Genes with Similar Biological Functions

Hui Wen Nies [1], Zalmiyah Zakaria [1], Mohd Saberi Mohamad [2,*], Weng Howe Chan [1], Nazar Zaki [3], Richard O. Sinnott [4], Suhaimi Napis [5], Pablo Chamoso [6], Sigeru Omatu [7] and Juan Manuel Corchado [6]

[1] School of Computing, Faculty of Engineering, Universiti Teknologi Malaysia, Skudai 81310, Johor, Malaysia
[2] Institute for Artificial Intelligence and Big Data, Universiti Malaysia Kelantan,
     Kota Bharu 16100, Kelantan, Malaysia
[3] Department of Computer Science and Software Engineering, College of Information Technology,
     United Arab Emirate University, Al Ain 15551, UAE
[4] School of Computing and Information Systems, University of Melbourne, Parkville 3010, Victoria, Australia
[5] Faculty of Biotechnology and Biomolecular Sciences, Universiti Putra Malaysia,
     Serdang 43400, Selangor, Malaysia
[6] BISITE Research Group, Digital Innovation Hub, University of Salamanca, Edificio I+D+i, C/ Espejos s/n,
     37007 Salamanca, Spain
[7] Division of Data-Driven Smart Systems Design, Digital Monozukuri (Manufacturing) Education and
     Research Center, Hiroshima University, #210, 3-10-31 Kagamiyama, Higashi-Hiroshima 739-0046,
     Hiroshima Prefecture, Japan
*   Correspondence: saberi@umk.edu.my

Received: 8 July 2019; Accepted: 16 August 2019; Published: 21 August 2019

Abstract: Clustering techniques can group genes based on similarity in biological functions. However, the drawback of using clustering techniques is the inability to identify an optimal number of potential clusters beforehand. Several existing optimization techniques can address the issue. Besides, clustering validation can predict the possible number of potential clusters and hence increase the chances of identifying biologically informative genes. This paper reviews and provides examples of existing methods for clustering genes, optimization of the objective function, and clustering validation. Clustering techniques can be categorized into partitioning, hierarchical, grid-based, and density-based techniques. We also highlight the advantages and the disadvantages of each category. To optimize the objective function, here we introduce the swarm intelligence technique and compare the performances of other methods. Moreover, we discuss the differences of measurements between internal and external criteria to validate a cluster quality. We also investigate the performance of several clustering techniques by applying them on a leukemia dataset. The results show that grid-based clustering techniques provide better classification accuracy; however, partitioning clustering techniques are superior in identifying prognostic markers of leukemia. Therefore, this review suggests combining clustering techniques such as CLIQUE and k-means to yield high-quality gene clusters.

Keywords: gene clustering; swarm intelligence; biological functions detection; informative genes

---

1. Introduction

Analysis of gene expression levels is essential in studying and detecting genes functions. According to Chandra and Tripathi [1], genes that have similar gene expression levels are likely to involve similar biological functions. The authors showed that the clustering process was quite useful to identify co-expressed genes in a group of genes and, in addition, to detect unique genes in different groups. Therefore, clustering can be quite helpful to extract valuable knowledge from a large amount of biological data [2], which could lead to prevention, prognosis, and treatment in biomedical research.

Cai et al. [3] developed a random walk-based technique to cluster similar genes. The authors show that the proposed method was useful in strengthening the interaction between genes by considering the types of interactions that exist in the same group of genes. Many previous random walk-based methods managed to extract local information from a large graph without knowledge of the whole graph data [4]. In a random walk-based method, a gene is important if it interacts with many other genes [5–8]. As illustrated in Figure 1, gene 1 has a higher degree than gene 2 (two outgoing links) compared to one outgoing link from gene 3 to gene 4. In this case, gene 1 is the most important gene among the four genes shown in the hypothetical gene network.

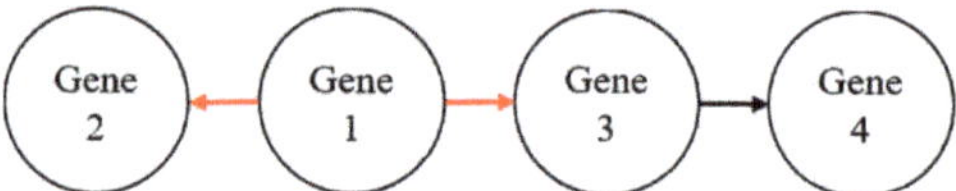

Figure 1. A hypothetical gene network to illustrate the importance of genes in a random walk.

Several previous studies have noted the importance of clustering to identify co-expressed genes in a cluster and inactive genes in another cluster [1,9]. Clustering can also discover the fundamental hidden structure of biomedical data, which can be used for diagnosis and treatments [9]. In addition, clustering is extremely vital for identifying cancer subtyping and the detection of the tumor.

Researchers typically focus on clustering by assuming the number of clusters beforehand, which can be seen in [10,11]. This problem can lead to the inability of the clustering techniques to obtain an optimal number of centroids and hence results in poor quality of clusters [11,12]. In previous studies, several proposed approaches managed to discover the optimal number of clusters by simply tuning and optimizing the parameters of the clustering method. This can be done by repeating the process of analyzing the eigenvalues of the affinity matrix, which are equal to the number of desired clusters [13]. In addition, rotating normalized eigenvectors and squared-loss mutual information (SMI) can be employed in the clustering process to obtain an optimal number of clusters [14,15]. Besides, the elbow method and the average silhouette method are the other examples to identify the optimal number of clusters in previous studies [15,16]. The elbow method identifies the optimal number of clusters by calculating sum of squared error for each number of clusters (k) from a range of k values. The average silhouette method computes the average silhouette values of genes for different values of k (number of clusters). Then, this method selects the optimal number of clusters that has the maximum average silhouette values from the range of k values. Optimization of the objective function and validation of clustering can improve the quality of clusters [11]. The optimization for the objective function of clustering can identify the best solution among a set of solutions. On the other hand, clustering validation is used to determine clusters in the data using an appropriate measurement [17]. Clustering validation can also evaluate the goodness of the clustering structure based on the given class labels [18]. Thus, validation is an essential step because it assists in the identification of which cluster is more informative compared to other clusters [19].

This paper focuses on reviewing existing computational methods on genes clustering using the notion of optimizing the objective function and validation.

2. Gene Network Clustering Techniques

In general, clustering can be categorized into partitioning, hierarchical, grid-based, and density-based techniques [11,17,20–22]. In Table 1, we show differences among categories of clustering techniques. The table also provides some information such as time complexity, computing efficiency, convergence rate, scalability, and initialization of cluster number. Partitioning clustering assigns the data objects into a number of clusters fixed beforehand. This technique identifies the number of centroids and assigns the objects to the nearest centroid. Hierarchical clustering groups the data based on the distance of the objects to form clusters. This technique can be either started with large data and aggregated into a small group or started from a small group of data and merged until all the data are in one large group. Grid-based clustering divides each dimension of data space to form a grid structure. Density-based clustering separates the data according to the density of the objects. Traditionally, hierarchical, grid-based, and density-based techniques do not require cluster number as an input parameter [20,23]. In the view of Jain [17], hierarchical clustering is more versatile than partitioning clustering. With the discovery of clusters with good robustness and flexibility, grid-based and density-based techniques have been particularly useful [24]. They are also helpful for dealing with large spatial data and the proper use of expert knowledge. Grid-based and density-based techniques also aim to identify data densities and to split the data space into grid structures when looking for groupings [25]. Grid-based clustering techniques are more efficient compared to density-based clustering techniques; however, the use of summarized information makes these techniques lose effectiveness in cases where the number of dimensions increases [26].

Table 1. Differences among categories of clustering techniques.

Categories	Time Complexity	Computing Efficiency	Convergence Rate	Scalability	Initialization of Cluster Number
Partitioning	Low	High	Low	Low	Yes
Hierarchical	High	High	Low	High	No
Grid-based	Low	High	Low	High	No
Density-based	Middle	High	High	High	No

In Table 2, we present several examples of clustering techniques done by previous researchers. The table also summarizes the advantages and the disadvantages of the techniques. From this table, k-means clustering is the most popular technique, even though k-means suffers from the shortcoming of identifying the number of potential clusters before the clustering setup.

Table 2. Examples of popular clustering techniques along with their advantages and disadvantages.

Clustering Techniques	Categories	Advantages	Disadvantages	References
Fuzzy C Means (FCM)	Partitioning	Minimize the error function belonging to its objective function and solve the partition factor of the classes.	Unable to achieve high convergence.	[27,28]
K-means Clustering	Partitioning	Use a minimum "within-class sum of squares from the centers" criterion to select the clusters.	Need to initialize the number of clusters beforehand.	[9–12,29–33]
Partitioning Around Medoids (PAM)	Partitioning	Deal with interval-scaled measurements and general dissimilarity coefficients.	Consumes large central memory size.	[34]
Self-Organizing Maps (SOM)s	Partitioning	Suitable for data survey and getting good insight into the cluster structure of data for data mining purposes.	Distance dissimilarity is ignored.	[35–38]
Agglomerative Nesting (AGNES)	Hierarchical (agglomerative)	Build a hierarchy of clustering from a small cluster and then merge until all data are in one large group.	Starts with details and then works up to large clusters, which is affected by unfortunate decisions in the first step.	[19,34]
EISEN Clustering	Hierarchical (agglomerative)	Carry out a clustering in which a mean vector represents each cluster from data in the group.	Starts with details and then works up to large clusters, which can be affected by unfortunate decisions in the first step.	[19]
Divisive Analysis (DIANA)	Hierarchical (divisive)	Perform a task starting from a large cluster containing all data to only a single dataset.	Not generally available and rarely applied in most studies.	[19,34]
Clustering in Quest (CLIQUE)	Grid-based	Can automatically find subspaces in lower-dimensional subspaces with high-density clusters.	Ignores all projections of dimensional subspaces.	[39,40]

Table 2. *Cont.*

Clustering Techniques	Categories	Advantages	Disadvantages	References
Grid-Clustering Technique for High-Dimensional and Large Spatial Databases (GCHL)	Grid-based	Efficient and scalable while handling high dimensionality issue.	Insensitive to noise.	[26,41]
Statistical Information Grid (STING)	Grid-based	Facilitate several kinds of spatial queries and less computational cost.	Difficult to identify multiple clusters.	[42,43]
Density-Based Spatial Clustering of Applications with Noise (DBSCAN)	Density-based	Can detect clusters with different shapes and able to handle ones with different densities.	Optimization issue. Difficult to select appropriate parameter values.	[44,45]
Random Walk based Clustering	Density-based	Reflect the topological features of a functional network.	Considers the interaction between two genes.	[46–48]
Relative Core Merge (RECOME)	Density-based	Can characterize based on a step function of its parameter.	Scalability issue. Hard to handle a large volume of data.	[45]

According to the reviewed clustering techniques in Table 2, this experimental work aims to investigate which category of clustering techniques would perform better in clustering genes. Gene expression data from the leukemia microarray study by Golub et al. [49] are used in this study. These data consist of 3051 genes, 38 tumor mRNA samples [27 acute lymphoblastic leukemia (ALL) and 11 acute myeloid leukemia (AML)] [50]. The clustering techniques investigated in this experimental work are k-means clustering (partitioning), agglomerative nesting (AGNES) (hierarchical), clustering in quest (CLIQUE) (grid-based), and density-based spatial clustering of applications with noise (DBSCAN) (density-based). The results in terms of percentage of accuracy are shown in Table 3. The experimental work was carried out using stratified ten-fold cross-validation and a support vector machine as a classifier. The selected clusters in Table 3 were validated based on silhouette width. According to Table 3, the CLIQUE was able to achieve the highest classification accuracy when applied on the leukemia dataset compared to other clustering techniques. In addition, Table 3 also shows several genes were biologically validated as prognostic markers for leukemia when PubMed text mining was used. Prognostic marker was commonly used to differentiate between good or poor disease outcomes [51]. This validation was done to show the relationship between genes and prognostic markers of leukemia [52]. Although CLIQUE achieved the best classification accuracy, the technique identified 67 genes as prognostic markers of leukemia out of 919 genes in the selected cluster. On the other hand, k-means had the best performance in identifying prognostic markers of leukemia (8%). The remaining techniques were able to achieve between 6% and 8% in determining the prognostic markers of leukemia over the number of genes in the selected clusters.

Table 3. Comparative results of the clustering technique applied on leukemia gene expression data.

Categories	Clustering Techniques	Parameter (s)	Number of Genes in the Selected Cluster	Number of Prognostic Markers	Accuracy (%)
Partitioning	K-means	k = 2	275	22	71.50
Hierarchical	AGNES	k = 2	339	22	78.50
Grid-based	CLIQUE	k = 2 dimension = 10 density = 0.2	919	67	89.00
Density-based	DBSCAN	k = 2 minPts = 10	1548	103	73.00

Note: k is the number of clusters to be selected; dimensions are divided into several equal-width intervals; density is the density threshold; minPts is the minimum size of clusters.

2.1. Category 1: Partitioning Clustering

Detection of clusters using partitioning clustering has low time complexity and high computational cost [53]. However, there are specific problems related to this technique. One of these problems is detecting clusters inappropriate for non-convex data. This could be because clustering techniques cannot spatially separate the data [54]. Other disadvantages are the need to initialize the number of clusters beforehand, and that the clustering result is sensitive to the intended number of possible clusters. Fuzzy C Means (FCM), k-means clustering, Partitioning Around Medoids (PAM), and Self-Organizing Maps (SOM) are all examples of partitioning clustering [9–12,27–38]. PAM is a variation of k-means clustering [55], and it is more robust in terms of accuracy compared to k-means clustering, for instance, when applied to classify cancer types [56,57].

2.2. Category 2: Hierarchical Clustering

Hierarchical clustering's scalability is relatively high in cluster detection [53]. One benefit of the method is that it can detect the hierarchical relationship among clusters easily. However, the major drawback associated with hierarchical clustering is the high computational cost. Agglomerative (bottom-up) and divisive (top-down) are the categories of hierarchical clustering [2,35,58]. The way of merging clusters and identification of the node levels can differentiate between agglomerative and divisive hierarchical clustering [58]. Agglomerative hierarchical clustering (AHC) combines the most adjacent pair of clusters, forming a group from bottom to top [59]. Several strategies of AHC are used to identify the distance between clusters, which are single linkage, complete linkage, centroid linkage, average linkage, Ward's method, and the probability-based method [25,58,59]. On the other hand, divisive hierarchical clustering is useful to identify clusters with different densities and shapes [58,59]. The method starts from all samples in a group and then splits the samples into two sub-clusters, which are then divided into further sub-clusters and so on [58]. For AHC, node-level is the diameter of a new cluster formed at the splitting step. The node-level of divisive methods is to divide the groups based on their diameters. Agglomerative nesting (AGNES), EISEN clustering, and divisive analysis (DIANA) are examples of hierarchical clustering [19,34]. Garzón and González [19] used these clustering techniques to group similar genes before the step of the gene selection.

2.3. Category 3: Grid-Based Clustering

The design of grid-based clustering divides the entire data space into multiple, non-overlapping grid structures [24,59]. This method performs faster than density-based clustering. Grid-based clustering can benefit from dividing the data space into grids to reduce its time complexity [22,60]. CLIQUE, grid-clustering technique for high-dimensional very large spatial databases (GCHL), and statistical information grid (STING) are examples of grid-based clustering [39–43]. The GCHL technique can discover concave (deeper) and convex (higher) regions when applied in medical and geographical fields and by using the average eight direction (AED) technique [26,41]. However, both techniques struggle to identify complex clusters from high dimensional data. CLIQUE partitions the data space into cells and searches subspaces by counting the number of points in each cell [61]. Searching a suitable set of dimensions for each cluster can form the candidate subspace for the centroid of the cluster. Different groups of points are clustered in different subspaces [62].

2.4. Category 4: Density-Based Clustering

Usually, the regions contain points with high density in the data space, which makes density-based clustering mistake them as clusters [59]. Mechanisms of aggregation in density can characterize the clustering [45]. A significant advantage of density-based clustering is that it can discover differently shaped clusters and noise from data [22,24,63]. However, density-based clustering has a high runtime analysis to detect clusters [64]. DBSCAN, random walk, and Relative Core Merge (RECOME) are examples of density-based clustering [44–48]. Historically, a random walk uses the theory of Markov chain [48,65]. In most studies, the random walk has been used to infer and to optimize the structural properties of networks [65,66]. Much of the current literature on the random walk is on ranking the genes concerning their specific probabilities from high to low [67,68]. In literature, a random walk mostly uses the topological similarity in networks to identify genes with a similar disease.

3. Optimization for Objective Function of Partitioning Clustering Techniques

Optimization for objective function can improve the efficiency of partitioning clustering techniques during initialization of the intended cluster number [11,33]. Swarm intelligence is widely used as the objective function for a clustering problem. The number of intended clusters can be predicted based on the typical search of the patterns [69,70]. Swarm intelligence can also be applied through maximizing or minimizing the objective function of clustering [69,71,72]. In most studies, swarm intelligence has been mostly used in the field of optimization [73,74].

Swarm intelligence refers to the collective behavior of decentralized, self-organized systems of living creatures. The swarm intelligence systems consist typically of a population of simple agents or boids interacting locally with one another and with their environment. The inspiration often comes from nature, especially biological systems [75,76].

For modeling the behavior of a swarm, the techniques are made up of animals and insects, such as bees, ants, birds, fishes, and so on [74,77]. Most recent studies used swarm intelligence to solve problematic real-world problems such as networking, traffic routing, robotics, economics, industry, games, etc. [73,74]. Hence, clustering techniques can benefit from swarm intelligence [74].

Swarm intelligence can optimize the objective function of clustering based on population and evolution strategies [11,33]. This function is usually used to determine the fitness of each particle since the community has a set of particles (known as a swarm), and each particle represents a solution. Table 4 compares the use of optimization in population and evolution strategies. Both optimization strategies are designed to imitate the best features in nature and produce a better quality of solution efficiently [78,79]. Previous studies have explored the use of optimization in a generation with more than 1000 populations before the convergence step, but it was not computationally efficient [80].

Table 4. Comparison of the use of optimization between population and evolution strategies.

Strategies	Population-Based		Evolution
Functions	**Exploration**	**Exploitation**	
Between technique and solution	The technique can reach the best solution within the search space.	Express the ability of the technique to reach the global optimum solution, which was around the obtained local solutions.	Optimize the mathematical functions of the technique with continuously changeable parameters and extend to solve discrete optimization problems.
Application	Metaheuristic search for global optimal solutions using informative parameters.		Processes of selection, recombination, and mutation.
Weakness	Difficult to avoid problems of local minima and early convergence.		Need to control and adjust parameters.
Aim	Imitate the best features in nature and produce a better quality of solution efficiently.		

Table 5 summarizes existing techniques of swarm intelligence based on the strategies together with their usages. Xu et al. [81] found particle swarm optimization (PSO) is faster than both artificial bee colony (ABC) and genetic algorithm (GA) because PSO can perform without any complicated evolution. Previous studies have also shown some drawbacks of ABC, which are the limited ability of exploitation, slow convergence speed, and low-quality solutions [82]. In the review of GA and PSO algorithms, Gandomi et al. [79] identified the main purposes of these techniques, which solved significant problems faster.

Table 5. Summary of existing techniques of swarm intelligence.

Techniques	Strategies	Usage	Fitness	References
Artificial Bee Colony (ABC)	Population	Can stimulate searching food process of bees based on the found food sources quality.	Position and nectar amount of a food source.	[37,82]
Ant Colony Optimization (ACO)	Population	Mimic ant behavior to solve optimization problems.	Pheromone values.	[77,83]
Ant Lion Optimization (ALO)	Population	High exploitation to explore search space and quickly converge to a global optimum.	Ant location.	[11,33]
Bat Algorithm	Population	Uses the frequency-based tuning and pulse emission rate changes that can lead to better convergence.	Bat behavior.	[78,80,84]
Bee Algorithm	Population	Imitate food foraging behavior of swarms of honeybees to find the optimal solution.	Frequency of the dance.	[85]
Cuckoo Search (CS)	Population	Combine the obligate brood parasitic behavior of some cuckoo species with Lévy flight behavior of some birds and fruit flies.	Quality of cuckoo bird eggs.	[79]
Firefly Algorithm (FA)	Population	Carry out nonlinear design optimization and solve unconstrained stochastic functions.	Brightness of the firefly.	[70,86]
Gravitational Search Algorithm (GSA)	Population	Emulate the law of Newtonian gravity to solve various nonlinear optimization problems.	Intelligence factors.	[87,88]
Particle Swarm Optimization (PSO)	Population	Balance the weights of a neural network and sweep the search space using a swarm of particles.	A "space" where the particles "move".	[71,77,81,89]
Simulated Annealing (SA)	Population	Use principles of statistical mechanics regarding the behavior of many atoms at low temperature.	Single bit-flips.	[90,91]
Differential Evolution (DE)	Evolution	Maintain a population of target vectors at each iteration for stochastic search and global optimization.	Global minimum.	[71]
Evolution Strategy (ES)	Evolution	Emphasize the use of normally distributed random mutations (main operator).	Several operators needed to consider in the analysis.	[92]
Evolutionary Programming (EP)	Evolution	Use the self-adaptation principle to evolve the parameters on searching.	No recombination operator and difficult to identify useful values for parameter tuning.	[92]
Gene Expression Programming (GEP)	Evolution	Extremely versatile and greatly surpasses the existing evolutionary techniques.	Several genetic operators needed to function on selected chromosomes during reproduction.	[93,94]
Genetic Algorithm (GA)	Evolution	Use genes with mechanisms to mimic survival of the fittest and inspire the genetics with the evolution of populations.	Priority of the genetic strings.	[71]
Genetic Programming (GP)	Evolution	Can select variables and operators automatically then assemble into suitable structures.	No clearly defined termination point in biological processes operating.	[95,96]
Memetic Algorithm	Evolution	Useful on the property of global convexity in the search space.	Genetic operators (*crossover* and *mutation*) needed to consider in the analysis.	[97–99]

3.1. Strategy 1: Population-Based Optimization

Population-based optimization is performed in terms of exploration and exploitation [69,100]. Exploration is the technique able to reach the best solution within the search space, while exploitation expresses the ability of the technique to reach a global optimum solution. Metaheuristic search can apply in this optimization for global optimal solutions using informative parameters. However, the optimization still has difficultly avoiding the problems of local minima and early convergence [11,33,101]. Several examples of population-based optimization are reviewed, which are ant colony optimization (ACO), ant lion optimization (ALO), firefly algorithm (FA), and particle swarm optimization (PSO) [11,33,70,71,77,81,83,86,89].

In the literature related to PSO, most previous studies used PSO because it does not have any complex evolution [81]. Fister et al. [70] found that FA is suitable for multi-modal optimization and fast convergence.

3.2. Strategy 2: Evolution-Based Optimization

Evolution-based optimization is involved in the processes of selection, recombination, and mutation [102]. The selection of evolution strategy fails to deal with changing environments, and it threatens the self-adaptation with its control parameters (internal model) [103,104]. For recombination processes (in terms of discrete and intermediate processes), it performs with control parameters on object variables, standard deviations, and rotation angles. The mutation mechanism makes the techniques evolve its control parameters (standard deviations and covariances). Evolution-based optimization can optimize the mathematical functions of the technique with continuously changeable parameters and extend to solve discrete optimization problems. This strategy can deliver a high quality of solutions and allows the technique to move toward better solutions in the search space with a population [105,106]. GA is one of the techniques using evolution strategy, which is commonly used for clustering based on selection, crossover, and mutation. In previous studies, most algorithms were derived from GA, such as evolution strategy (ES) and evolutionary programming (EP) [92]. The memetic algorithm is the extension of GA and includes local search optimization for problem-solving [97–99]. Genetic programming (GP), on the other hand, is the extension of GA that has been successfully applied and used to solve many problems [95,96]. Moreover, gene expression programming (GEP) uses the character of linear chromosomes and has been applied in symbolic regression and block stacking [93,94].

4. Clustering Validation in Measurements

Previous studies have evaluated the identified gene clustering in terms of distance [1]. If they are not within a distance regarding a specified gene in each experimental condition, then the specified gene is classified as an inactive gene. Otherwise, the specified gene is co-expressed.

Clustering validation can be measured in terms of internal and external criteria [17,18,100,107]. Table 6 summarizes the differences between internal and external validations. In general, internal criteria can assess the fitness between clustering structure and data. External criteria can measure the performance by matching cluster structure to prior information. As mentioned by Handl et al. [23], internal validation suffers from bias regarding clusters number and partitioning structure from data. The goal of internal validation is measured based on compactness and separation [18,107]. Compactness is defined as a measure of how close the objects are in a cluster based on variance. Separation measures either how a cluster is distinct or how well separated it is from other clusters. Handl et al. [23] held the view that external validation can suffer from biases in a partitioning according to cluster number and distribution of groups with class sizes.

Table 7 sets out examples of measurements to validate the quality of clusters. As can be seen from the table, previous studies commonly used Euclidean distance and silhouette width. In general, silhouette width can validate the clustering performance in terms of pairwise difference between and within cluster distances [18,107]. The maximum values of the silhouette width can identify an optimal number of clusters.

Table 6. The difference in measurements between internal and external validations.

Criteria of Validation Measurements	Internal	External
Aim	Assess the fitness between clustering structure and data.	Measure the performance by matching cluster structure to prior information.
Suffer from bias	• Number of clusters • Partitioning structure from data	• Number of clusters • Distribution of cluster with class sizes in a partitioning

Table 7. Examples of previous studies in clustering validation.

Measurements	Categories	Usage	References
Average of sum of intra-cluster distances	Internal	Measure assessing cluster compactness or homogeneity.	[11,33]
Connectivity	Internal	Degree of the connectedness of clusters.	[1,23]
Davies and Bouldin (DB) index	Internal	Measure intra- and inter-cluster using spatial dissimilarity function.	[108]
Dunn index	Internal	Ratio of the smallest distance among observations in the different cluster to the most considerable intra-cluster distance.	[1,23]
Euclidean distance	Internal	Compute distances between the objects to quantify their degree of dissimilarity.	[19,31,34,109]
Inter-cluster distance	Internal	Quantify the degree of separation between individual clusters.	[11]
Manhattan distance	Internal	Correspond to the sum of lengths of the other two sides of a triangle.	[34]
Pearson correlation coefficients (PCC)	Internal	Measure between-state functional similarity.	[23,110]
Silhouette width	Internal	Measure the degree of confidence in a clustering assignment and lie in the interval [−1, +1], with well-clustered observations having values near +1 and near -1 for poorly clustered observations.	[1,18,19,31,32,109]
Square sum function of the error	Internal	Measure the quality of cluster either by compactness or homogeneity.	[12,23,111]
Entropy	External	Measure mutual information based on the probability distribution of random variables.	[30,112,113]
F-measure	External	Assess the quality of clustering result at the level of entire partitioning and not for an individual cluster only.	[11,23,30,33]

5. Discussion

An efficient clustering technique is the one capable of extracting useful information about the behavior of a gene. According to Oyelade et al. [114], ensemble clustering (a combination of two or more phases of clustering) can generate more robust and better quality clusters compared to single clustering. Table 8 summarizes the ensemble methods for clustering that were used by previous researchers. In addition, Oyelade et al. [114] also showed that hierarchical clustering is more suitable to handle real datasets, such as image data, compared to partitioning clustering, but it is computationally expensive. Advanced technological developments can isolate a large group of cells. Biological data can provide a better understanding of the complex biological processes. For example, single-cell RNA sequencing can help to expose biological processes and medical insights [115]. The k-means clustering typically performs better than hierarchical clustering in smaller datasets, but it requires a long computational time [114,115]. Other than that, large amounts of bulk data can address biological dynamics and cancer heterogeneity. Tang et al. [115] proposed High-order Correlation Integration (HCI), which uses k-means clustering and Pearson's correlation coefficient in the experiments. Their results showed that HCI outperforms the existing methods (k-means clustering and hierarchical clustering) under single-cell and bulk RNA-seq datasets. Unsupervised clustering is one of the powerful techniques used in single-cell RNA sequencing to define cell types based on the transcriptome [116]. Fully unsupervised clustering techniques (e.g., intelligent k-means and kernel k-means) are applied to analyze genes in colorectal carcinoma [117]. Other than that, random walk-based clustering, GCHL, and CLIQUE clustering techniques are also used in unsupervised manners [26,41,46–48,61,67].

The purpose of optimization for objective function and validation is to achieve quality clusters. Most of the previous studies used swarm intelligence to optimize the parameters of clustering techniques and to identify the optimal number of possible clusters [118]. The objective function of clustering techniques defines optimization as maximizing the accuracy of the centroid or the cluster center, especially for partitioning clustering techniques. It is because partitioning clustering needs to initialize either the number of clusters or the number of centroids beforehand. Furthermore, clustering validation is also essential to measure within or between the identified clusters [19].

Table 8. Summary of the existing ensemble methods used in clustering.

References	Ensemble Methods	Clustering Techniques	Use
Deng et al. [24]	Grid-based and Density-based Spatial Clustering (GRIDEN)	Grid-based Density-based (DBSCAN)	Enhances clustering speed.
Oyelade et al. [114] Masciari et al. [119]	Microarray Data Clustering using Binary Splitting (M-CLUBS)	Hierarchical (divisive and agglomerative)	Overcomes the effect of size and shape of clusters, number of clusters, and noise for gene expression data.
Oyelade et al. [114] Bouguettaya et al. [120]	Efficient Agglomerative Hierarchical Clustering (KnA)	Hierarchical (agglomerative) Partitioning (k-means)	Relatively consistent in synthetic data.
Bouguettaya et al. [120] Lin et al. [121]	Cohesion-based Self-Merging (CSM)	Partitioning (k-means) Hierarchical (divisive)	Clusters the datasets of arbitrary shapes very efficiently.
Darong and Peng [122]	Grid-based DBSCAN Technique with Referential Parameters (GRPDBSCAN)	Grid-based Density-based (DBSCAN)	Finds clusters of arbitrary shape and removes noise.

In this research, leukemia data containing 3051 genes and 38 samples [49] were used to evaluate the performance of each clustering techniques category. The genes obtained by the clustering techniques were different from one technique category to another; however, the number of target clusters was the same among the techniques. As a result, the grid-based clustering technique provided higher classification accuracy than other clustering techniques. The technique was able to identify 7.29% of the prognostic markers in leukemia data. On the other hand, k-means clustering achieved the highest percentage (8%) of identifying prognostic markers in leukemia, but the classification accuracy in this case was quite poor.

A summary of optimal cluster analysis studied by previous researchers is shown in Table 9. According to the table, k-means clustering was the most used in the research. Integration of optimization is critical to its use in research because it can solve the issue of k-means clustering that requires initializing the number of clusters beforehand [10,11].

Table 9. Summary of optimal cluster analysis.

References	Clustering Techniques	Optimization for Objective Function of Partitioning Clustering Techniques	Clustering Validation
Majhi and Biswal [11,33]	K-means clustering	Ant Lion Optimization (ALO)	• Average of sum of intra-cluster distances • F-measure
Ye et al. [12]	K-means clustering	Cuckoo Search	Square sum function of the error
Mary and Raja [30]	K-means clustering	Ant Colony Optimization (ACO)	• Entropy • F-measure
Garg and Batra [32]	• Decision Tree Criterion (DTC) • K-means clustering	Cuckoo Search Optimization (CSO)	• Mean Square Error (MSE) • Silhouette width
Acharya et al. [91]	Multi-Objective Based Bi-Clustering	Simulated Annealing (SA)	Euclidean distance
Labed et al. [111]	K-Harmonic Means	Cuckoo Search Algorithm (CSA)	• Davies and Bouldin (DB) index • Square sum function of the error
Shanmugam and Sekaran [118]	Fuzzy C Means (FCM)	Ant Lion Optimization (ALO)	Square sum function of the error
Carneiro et al. [122]	Network-based techniques (e.g., clustering and dimensionality reduction)	Particle Swarm Optimization (PSO)	Euclidean distance

6. Conclusions

In summary, this paper reviewed examples of existing computational methods for clustering genes with similar biological functions. As a result, we found that partitioning, hierarchical, grid-based, and density-based are the categories of clustering techniques. Clustering can identify a high-quality cluster that is helpful in biological mechanisms and could lead to the identification of new genes related to potentially known or suspected cancer genes [67,117,123].

Among the categories of clustering, grid-based and density-based techniques are more suitable to be used to cluster objects in large spatial data. These techniques are inappropriate for artificial and biological datasets such as iris, wine, breast tissue, blood transfusion, and yeast datasets [24,114]. On the other hand, density-based clustering techniques are useful if used to cluster gene expression data [114]. Moreover, hierarchical clustering techniques are useful to handle synthetic and real datasets (e.g., image data). However, these techniques have some limitations when the data are very large [114]. Finally, partitioning clustering techniques are inappropriate for non-convex data but suitable for smaller datasets [53,114,115].

Grid-based clustering (CLIQUE) was more efficient than other categories of clustering (e.g., k-means clustering, DBSCAN, and AGNES), but it was difficult to identify multiple clusters in cases of high dimensional data types. Although k-means clustering (category: partitioning) was sensitive to initializing the number of clusters, it provided a higher chance of identifying prognostic markers of leukemia. A prognostic marker is useful for identifying a disease outcome, which can be helpful in cancer treatment and drug discovery as well [52]. However, the quality of clusters is usually affected by initializing the number of intended clusters, especially for partitioning clustering. Therefore, the optimization of the objective function and validation can help clustering techniques to identify the optimal number of clusters with better quality [11,89]. This paper also showed the two types of optimization strategies, which are population and evolution. Most of the existing techniques used for optimization utilize population strategies. Carneiro et al. [124] also concluded that the use of optimization could generate better classification together with the use of clustering and topological data. In addition, this paper also reviewed clustering validation and its measurements criteria. Internal and external criteria are commonly used to measure the cluster structure. Besides, genes in clusters can belong to a specific pathway, which can reflect the genes' functioning in biological processes [125]. For example, BCL2 associated with X apoptosis regulator (BAX) was among the genes identified in our experimental work, which is also a prognostic marker of leukemia. The BAX gene was encoded in the pro-apoptosis proteins, which could increase its expression and decrease the expression of anti-apoptosis (e.g., Bcl-2 gene) in the treatment of leukemia [126,127]. Moreover, clustered genes can identify metabolic gene clusters related to the discovery of metabolite in bacteria and fungi [127]. Identifying genes in clusters can not only allow us to discover the informative gene and the prognostic marker for the specific disease, but it can also provide a clue about the cluster dictated by signature enzymes. The signature enzyme can catalyze reactions and further tailor the product. Hence, the genes can be encoded in the pathway with enzymes.

Based on the experimental work, the CLIQUE and the k-means clustering techniques produce better results in terms of classification accuracy and identifying cancer markers. Therefore, this review suggests combining clustering techniques such as CLIQUE and k-means to yield more accurate gene clustering.

Although the optimal cluster analysis is the focus of this review, the findings can be applied to different areas.

Author Contributions: Conceptualization, H.W.N., Z.Z., M.S.M. and W.H.C.; Methodology, H.W.N., Z.Z., M.S.M., W.H.C. and N.Z.; Resources, H.W.N.; Writing—Original Draft Preparation, H.W.N.; Writing—Review and Editing, Z.Z., M.S.M., W.H.C., N.Z., R.O.S., S.N., P.C., S.O., J.M.C.; supervision, Z.Z., M.S.M. and W.H.C.

Funding: This research was funded by Fundamental Research Grant Scheme—Malaysia's Research Star Award (FRGS-MRSA) and Fundamental Research Grant Scheme (R.J130000.7828.4F973) from Ministry of Education Malaysia, ICT funding agency from United Arab Emirates University (G00001472), and Research University Grant from Universiti Teknologi Malaysia (Q.J130000.2628.14J68). The authors also would like to thank Universiti Teknologi Malaysia (UTM) for the support of UTM's Zamalah Scholarship.

Processes **2019**, 7, 550

Acknowledgments: The authors acknowledge support from the Ministry of Education Malaysia, United Arab Emirates University (UAEU), University of Salamanca (USAL), and Universiti Teknologi Malaysia (UTM).

Conflicts of Interest: The authors declare no conflict of interest.

References

1. Chandra, G.; Tripathi, S. A Column-Wise Distance-Based Approach for Clustering of Gene Expression Data with Detection of Functionally Inactive Genes and Noise. In *Advances in Intelligent Computing*; Springer: Singapore, 2019; pp. 125–149.
2. Xu, R.; Wunsch, D.C. Clustering algorithms in biomedical research: A review. *IEEE Rev. Biomed. Eng.* **2010**, 3, 120–154. [CrossRef]
3. Cai, B.; Wang, H.; Zheng, H.; Wang, H. An improved random walk-based clustering algorithm for community detection in complex networks. In Proceedings of the International Conference on Systems, Man, and Cybernetics (SMC), Anchorage, AK, USA, 9–12 October 2011; pp. 2162–2167.
4. Zhang, H.; Raitoharju, J.; Kiranyaz, S.; Gabbouj, M. Limited random walk algorithm for big graph data clustering. *J. Big Data* **2016**, 3, 26. [CrossRef]
5. Liu, W.; Li, C.; Xu, Y.; Yang, H.; Yao, Q.; Han, J.; Shang, D.; Zhang, C.; Su, F.; Li, X.; et al. Topologically inferring risk-active pathways toward precise cancer classification by directed random walk. *Bioinformatics* **2013**, 29, 2169–2177. [CrossRef]
6. Liu, W.; Bai, X.; Liu, Y.; Wang, W.; Han, J.; Wang, Q.; Xu, Y.; Zhang, C.; Zhang, S.; Li, X.; et al. Topologically inferring pathway activity toward precise cancer classification via integrating genomic and metabolomic data: Prostate cancer as a case. *Sci. Rep.* **2015**, 5, 13192. [CrossRef]
7. Liu, W.; Wang, W.; Tian, G.; Xie, W.; Lei, L.; Liu, J.; Huang, W.; Xu, L.; Li, E. Topologically inferring pathway activity for precise survival outcome prediction: Breast cancer as a case. *Mol. Biosyst.* **2017**, 13, 537–548. [CrossRef]
8. Wang, W.; Liu, W. Integration of gene interaction information into a reweighted random survival forest approach for accurate survival prediction and survival biomarker discovery. *Sci. Rep.* **2018**, 8, 13202. [CrossRef]
9. Mehmood, R.; El-Ashram, S.; Bie, R.; Sun, Y. Effective cancer subtyping by employing density peaks clustering by using gene expression microarray. *Pers. Ubiquitous Comput.* **2018**, 22, 615–619. [CrossRef]
10. Bajo, J.; De Paz, J.F.; Rodríguez, S.; González, A. A new clustering algorithm applying a hierarchical method neural network. *Log. J. IGPL* **2010**, 19, 304–314. [CrossRef]
11. Majhi, S.K.; Biswal, S. A Hybrid Clustering Algorithm Based on K-means and Ant Lion Optimization. In *Emerging Technologies in Data Mining and Information Security*; Springer: Singapore, 2019; pp. 639–650.
12. Ye, S.; Huang, X.; Teng, Y.; Li, Y. K-means clustering algorithm based on improved Cuckoo search algorithm and its application. In Proceedings of the 3rd International Conference on Big Data Analysis (ICBDA), Shanghai, China, 9–12 March 2018; pp. 422–426.
13. Zelnik-Manor, L.; Perona, P. Self-tuning spectral clustering. In *Advances in Neural Information Processing Systems*; Neural Information Processing Systems Foundation, Inc. (NIPS): Vancouver, BC, Canada, 2005; pp. 1601–1608.
14. Sugiyama, M.; Yamada, M.; Kimura, M.; Hachiya, H. On Information-Maximization Clustering: Tuning Parameter Selection and Analytic Solution. In Proceedings of the 28th International Conference on Machine Learning, Bellevue, WA, USA, 28 June–2 July 2011; pp. 65–72.
15. Pollard, K.S.; Van Der Laan, M.J. A method to identify significant clusters in gene expression data. In *U.C. Berkeley Division of Biostatistics Working Paper Series*; Working Paper 107; Berkeley Electronic Press: Berkeley, CA, USA, 2002.
16. Bholowalia, P.; Kumar, A. EBK-means: A clustering technique based on elbow method and k-means in WSN. *Int. J. Comput. Appl.* **2014**, 105. [CrossRef]
17. Jain, A.K. Data clustering: 50 years beyond K-means. *Pattern Recognit. Lett.* **2010**, 31, 651–666. [CrossRef]
18. Liu, Y.; Li, Z.; Xiong, H.; Gao, X.; Wu, J. Understanding of internal clustering validation measures. In Proceedings of the 10th International Conference on Data Mining (ICDM), Sydney, Australia, 13–17 December 2010; pp. 911–916.
19. Garzón, J.A.C.; González, J.R. A gene selection approach based on clustering for classification tasks in colon cancer. *Adv. Distrib. Comput. Artif. Intell. J.* **2015**, 4, 1–10.

20. Kriegel, H.P.; Kröger, P.; Sander, J.; Zimek, A. Density-based clustering. *Wiley Interdiscip. Rev.* **2011**, *1*, 231–240. [CrossRef]

21. Nagpal, A.; Jatain, A.; Gaur, D. Review based on data clustering algorithms. In Proceedings of the Conference on Information & Communication Technologies, Thuckalay, Tamil Nadu, India, 11–12 April 2013; pp. 298–303.

22. Chen, Y.; Tang, S.; Bouguila, N.; Wang, C.; Du, J.; Li, H. A Fast Clustering Algorithm based on pruning unnecessary distance computations in DBSCAN for High-Dimensional Data. *Pattern Recognit.* **2018**, *83*, 375–387. [CrossRef]

23. Handl, J.; Knowles, J.; Kell, D.B. Computational cluster validation in post-genomic data analysis. *Bioinformatics* **2005**, *21*, 3201–3212. [CrossRef]

24. Deng, C.; Song, J.; Sun, R.; Cai, S.; Shi, Y. GRIDEN: An effective grid-based and density-based spatial clustering algorithm to support parallel computing. *Pattern Recognit. Lett.* **2018**, *109*, 81–88. [CrossRef]

25. Murtagh, F.; Contreras, P. Algorithms for hierarchical clustering: An overview. *Wiley Interdiscip. Rev. Data Min. Knowl. Discov.* **2012**, *2*, 86–97. [CrossRef]

26. Pilevar, A.H.; Sukumar, M. GCHL: A grid-clustering technique for high-dimensional very large spatial data bases. *Pattern Recognit. Lett.* **2005**, *26*, 999–1010. [CrossRef]

27. Dembele, D.; Kastner, P. Fuzzy C-means method for clustering microarray data. *Bioinformatics* **2003**, *19*, 973–980. [CrossRef] [PubMed]

28. Nayak, J.; Naik, B.; Behera, H.S. Fuzzy C-means (FCM) clustering algorithm: A decade review from 2000 to 2014. In *Computational Intelligence in Data Mining-Volume 2*; Springer: New Delhi, India, 2015; pp. 133–149.

29. Datta, S.; Datta, S. Methods for evaluating clustering algorithms for gene expression data using a reference set of functional classes. *BMC Bioinform.* **2006**, *7*, 397. [CrossRef] [PubMed]

30. Mary, C.; Raja, S.K. Refinement of Clusters from K-Means with Ant Colony Optimization. *J. Theor. Appl. Inf. Technol.* **2009**, *6*, 28–32.

31. Remli, M.A.; Daud, K.M.; Nies, H.W.; Mohamad, M.S.; Deris, S.; Omatu, S.; Kasim, S.; Sulong, G. K-Means Clustering with Infinite Feature Selection for Classification Tasks in Gene Expression Data. In Proceedings of the International Conference on Practical Applications of Computational Biology & Bioinformatics, Porto, Portugal, 21–23 June 2017; pp. 50–57.

32. Garg, S.; Batra, S. Fuzzified cuckoo based clustering technique for network anomaly detection. *Comput. Electr. Eng.* **2018**, *71*, 798–817. [CrossRef]

33. Majhi, S.K.; Biswal, S. Optimal cluster analysis using hybrid K-Means and Ant Lion Optimizer. *Karbala Int. J. Mod. Sci.* **2018**, *4*, 347–360. [CrossRef]

34. Kaufman, L.; Rousseeuw, P.J. *Finding Groups in Data: An Introduction to Cluster Analysis*; John Wiley & Sons: Hoboken, NJ, USA, 2009; Volume 344.

35. Vesanto, J.; Alhoniemi, E. Clustering of the self-organizing map. *IEEE Trans. Neural Netw.* **2000**, *11*, 586–600. [CrossRef] [PubMed]

36. Bassani, H.F.; Araujo, A.F. Dimension selective self-organizing maps with time-varying structure for subspace and projected clustering. *IEEE Trans. Neural Netw. Learn. Syst.* **2015**, *26*, 458–471. [CrossRef]

37. Mikaeil, R.; Haghshenas, S.S.; Hoseinie, S.H. Rock penetrability classification using artificial bee colony (ABC) algorithm and self-organizing map. *Geotech. Geol. Eng.* **2018**, *36*, 1309–1318. [CrossRef]

38. Tian, J.; Gu, M. Subspace Clustering Based on Self-organizing Map. In Proceedings of the 24th International Conference on Industrial Engineering and Engineering Management 2018, Changsha, China, 19–21 May 2018; pp. 151–159.

39. Agrawal, R.; Gehrke, J.; Gunopulos, D.; Raghavan, P. *Automatic Subspace Clustering of High Dimensional Data for Data Mining Applications*; ACM: New York, NY, USA, 1998; Volume 27, pp. 94–105.

40. Santhisree, K.; Damodaram, A. CLIQUE: Clustering based on density on web usage data: Experiments and test results. In Proceedings of the 3rd International Conference on Electronics Computer Technology (ICECT), Kanyakumari, India, 8–10 April 2011; Volume 4, pp. 233–236.

41. Cheng, W.; Wang, W.; Batista, S. Grid-based clustering. In *Data Clustering*; Chapman and Hall, CRC Press: London, UK, 2018; pp. 128–148.

42. Wang, W.; Yang, J.; Muntz, R. STING: A statistical information grid approach to spatial data mining. In Proceedings of the 23rd International Conference on Very Large Data Bases, Athens, Greece, 25–29 August 1997; Volume 97, pp. 186–195.

43. Hu, J.; Pei, J. Subspace multi-clustering: A review. *Knowl. Inf. Syst.* **2018**, *56*, 257–284. [CrossRef]

44. Ester, M.; Kriegel, H.P.; Sander, J.; Xu, X. A density-based algorithm for discovering clusters in large spatial databases with noise. In Proceedings of the Second International Conference on Knowledge Discovery and Data Mining, Portland, OR, USA, 2–4 August 1996; Volume 96, pp. 226–231.
45. Geng, Y.A.; Li, Q.; Zheng, R.; Zhuang, F.; He, R.; Xiong, N. RECOME: A new density-based clustering algorithm using relative KNN kernel density. *Inf. Sci.* **2018**, *436*, 13–30. [CrossRef]
46. Can, T.; Çamoğlu, O.; Singh, A.K. Analysis of protein-protein interaction networks using random walks. In Proceedings of the 5th International Workshop on Bioinformatics, Chicago, IL, USA, 21 August 2005; pp. 61–68.
47. Firat, A.; Chatterjee, S.; Yilmaz, M. Genetic clustering of social networks using random walks. *Comput. Stat. Data Anal.* **2007**, *51*, 6285–6294. [CrossRef]
48. Re, M.; Valentini, G. Random walking on functional interaction networks to rank genes involved in cancer. In Proceedings of the International Conference on Artificial Intelligence Applications and Innovations (IFIP), Halkidiki, Greece, 27–30 September 2012; pp. 66–75.
49. Golub, T.R.; Slonim, D.K.; Tamayo, P.; Huard, C.; Gaasenbeek, M.; Mesirov, J.P.; Coller, H.; Loh, M.L.; Downing, J.R.; Caligiuri, M.A.; et al. Molecular classification of cancer: Class discovery and class prediction by gene expression monitoring. *Science* **1999**, *286*, 531–537. [CrossRef] [PubMed]
50. Dudoit, S.; Fridlyand, J.; Speed, T.P. Comparison of discrimination methods for the classification of tumors using gene expression data. *J. Am. Stat. Assoc.* **2002**, *97*, 77–87. [CrossRef]
51. Ricci, C.; Marzocchi, C.; Battistini, S. MicroRNAs as biomarkers in amyotrophic lateral sclerosis. *Cells* **2018**, *7*, 219. [CrossRef] [PubMed]
52. Eyileten, C.; Wicik, Z.; De Rosa, S.; Mirowska-Guzel, D.; Soplinska, A.; Indolfi, C.; Jastrzebska-Kurkowska, I.; Czlonkowska, A.; Postula, M. MicroRNAs as Diagnostic and Prognostic Biomarkers in Ischemic Stroke—A Comprehensive Review and Bioinformatic Analysis. *Cells* **2018**, *7*, 249. [CrossRef] [PubMed]
53. Xu, D.; Tian, Y. A comprehensive survey of clustering algorithms. *Ann. Data Sci.* **2015**, *2*, 165–193. [CrossRef]
54. Halkidi, M.; Vazirgiannis, M. Clustering validity assessment: Finding the optimal partitioning of a data set. In Proceedings of the IEEE International Conference on Data Mining (ICDM), San Jose, CA, USA, 29 November–2 December 2001; pp. 187–194.
55. Rechkalov, T.V. Partition Around Medoids Clustering on the Intel Xeon Phi Many-Core Coprocessor. In Proceedings of the 1st Ural Workshop on Parallel, Distributed, and Cloud Computing for Young Scientists (Ural-PDC 2015), Yekaterinburg, Russia, 17 November 2015; Volume 1513.
56. Kumar, P.; Wasan, S.K. Comparative study of k-means, pam and rough k-means algorithms using cancer datasets. In Proceedings of the CSIT: 2009 International Symposium on Computing, Communication, and Control (ISCCC 2009), Singapore, 9 October 2011; Volume 1, pp. 136–140.
57. Mushtaq, H.; Khawaja, S.G.; Akram, M.U.; Yasin, A.; Muzammal, M.; Khalid, S.; Khan, S.A. A Parallel Architecture for the Partitioning around Medoids (PAM) Algorithm for Scalable Multi-Core Processor Implementation with Applications in Healthcare. *Sensors* **2018**, *18*, 4129. [CrossRef] [PubMed]
58. Roux, M. A Comparative Study of Divisive and Agglomerative Hierarchical Clustering Algorithms. *J. Classif.* **2018**, *35*, 345–366. [CrossRef]
59. Wang, J.; Zhu, C.; Zhou, Y.; Zhu, X.; Wang, Y.; Zhang, W. From Partition-Based Clustering to Density-Based Clustering: Fast Find Clusters with Diverse Shapes and Densities in Spatial Databases. *IEEE Access* **2018**, *6*, 1718–1729. [CrossRef]
60. Ding, F.; Wang, J.; Ge, J.; Li, W. Anomaly Detection in Large-Scale Trajectories Using Hybrid Grid-Based Hierarchical Clustering. *Int. J. Robot. Autom.* **2018**, *33*. [CrossRef]
61. Vijendra, S. Efficient clustering for high dimensional data: Subspace based clustering and density-based clustering. *Inf. Technol. J.* **2011**, *10*, 1092–1105. [CrossRef]
62. Yu, X.; Yu, G.; Wang, J. Clustering cancer gene expression data by projective clustering ensemble. *PLoS ONE* **2017**, *12*, e0171429. [CrossRef] [PubMed]
63. Bryant, A.; Cios, K. RNN-DBSCAN: A density-based clustering algorithm using reverse nearest neighbor density estimates. *IEEE Trans. Knowl. Data Eng.* **2018**, *30*, 1109–1121. [CrossRef]
64. Deng, C.; Song, J.; Sun, R.; Cai, S.; Shi, Y. Gridwave: A grid-based clustering algorithm for market transaction data based on spatial-temporal density-waves and synchronization. *Multimed. Tools Appl.* **2018**, *77*, 29623–29637. [CrossRef]

65. Pons, P.; Latapy, M. Computing communities in large networks using random walks. *J. Graph Algorithms Appl.* **2006**, *10*, 191–218. [CrossRef]

66. Petrochilos, D.; Shojaie, A.; Gennari, J.; Abernethy, N. Using random walks to identify cancer-associated modules in expression data. *BioData Min.* **2013**, *6*, 17. [CrossRef] [PubMed]

67. Ma, C.; Chen, Y.; Wilkins, D.; Chen, X.; Zhang, J. An unsupervised learning approach to find ovarian cancer genes through integration of biological data. *BMC Genom.* **2015**, *16*, S3. [CrossRef] [PubMed]

68. Zhu, L.; Su, F.; Xu, Y.; Zou, Q. Network-based method for mining novel HPV infection related genes using random walk with restart algorithm. *Biochim. Biophys. Acta Mol. Basis Dis.* **2018**, *1864*, 2376–2383. [CrossRef] [PubMed]

69. Civicioglu, P.; Besdok, E. A conceptual comparison of the Cuckoo-search, particle swarm optimization, differential evolution and artificial bee colony algorithms. *Artif. Intell. Rev.* **2013**, *39*, 315–346. [CrossRef]

70. Fister, I.; Fister, I., Jr.; Yang, X.S.; Brest, J. A comprehensive review of firefly algorithms. *Swarm Evol. Comput.* **2013**, *13*, 34–46. [CrossRef]

71. De Barros Franco, D.G.; Steiner, M.T.A. Clustering of solar energy facilities using a hybrid fuzzy c-means algorithm initialized by metaheuristics. *J. Clean. Prod.* **2018**, *191*, 445–457. [CrossRef]

72. Mortazavi, A.; Toğan, V.; Moloodpoor, M. Solution of structural and mathematical optimization problems using a new hybrid swarm intelligence optimization algorithm. *Adv. Eng. Softw.* **2019**, *127*, 106–123. [CrossRef]

73. Karaboga, D.; Akay, B. A survey: Algorithms simulating bee swarm intelligence. *Artif. Intell. Rev.* **2009**, *31*, 61–85. [CrossRef]

74. García, J.; Crawford, B.; Soto, R.; Astorga, G. A clustering algorithm applied to the binarization of Swarm intelligence continuous metaheuristics. *Swarm Evol. Comput.* **2019**, *44*, 646–664. [CrossRef]

75. Beni, G.; Wang, J. Swarm intelligence in cellular robotic systems. In *Robots and Biological Systems: Towards a New Bionics?* Springer: Berlin/Heidelberg, Germany, 1993; pp. 703–712.

76. Abraham, A.; Das, S.; Roy, S. Swarm intelligence algorithms for data clustering. In *Soft Computing for Knowledge Discovery and Data Mining*; Springer: Boston, MA, USA, 2008; pp. 279–313.

77. Pacheco, T.M.; Gonçalves, L.B.; Ströele, V.; Soares, S.S.R. An Ant Colony Optimization for Automatic Data Clustering Problem. In Proceedings of the IEEE Congress on Evolutionary Computation (CEC), Rio de Janeiro, Brazil, 8–13 July 2018; pp. 1–8.

78. Gandomi, A.H.; Yang, X.S.; Alavi, A.H.; Talatahari, S. Bat algorithm for constrained optimization tasks. *Neural Comput. Appl.* **2013**, *22*, 1239–1255. [CrossRef]

79. Gandomi, A.H.; Yang, X.S.; Alavi, A.H. Cuckoo search algorithm: A metaheuristic approach to solve structural optimization problems. *Eng. Comput.* **2013**, *29*, 17–35. [CrossRef]

80. Das, D.; Pratihar, D.K.; Roy, G.G.; Pal, A.R. Phenomenological model-based study on electron beam welding process, and input-output modeling using neural networks trained by back-propagation algorithm, genetic algorithms, particle swarm optimization algorithm and bat algorithm. *Appl. Intell.* **2018**, *48*, 2698–2718. [CrossRef]

81. Xu, X.; Li, J.; Zhou, M.; Xu, J.; Cao, J. Accelerated Two-Stage Particle Swarm Optimization for Clustering Not-Well-Separated Data. *IEEE Trans. Syst. Man Cybern. Syst.* **2018**, 1–12. [CrossRef]

82. Cao, Y.; Lu, Y.; Pan, X.; Sun, N. An improved global best guided artificial bee colony algorithm for continuous optimization problems. In *Cluster Computing*; Springer: Berlin, Germany, 2018; pp. 1–9.

83. Li, Y.; Wang, G.; Chen, H.; Shi, L.; Qin, L. An ant colony optimization-based dimension reduction method for high-dimensional datasets. *J. Bionic Eng.* **2013**, *10*, 231–241. [CrossRef]

84. Cheng, C.; Bao, C. A Kernelized Fuzzy C-means Clustering Algorithm based on Bat Algorithm. In Proceedings of the 2018 10th International Conference on Computer and Automation Engineering, Brisbane, Australia, 24–26 February 2018; pp. 1–5.

85. Ghaedi, A.M.; Ghaedi, M.; Vafaei, A.; Iravani, N.; Keshavarz, M.; Rad, M.; Tyagi, I.; Agarwal, S.; Gupta, V.K. Adsorption of copper (II) using modified activated carbon prepared from Pomegranate wood: Optimization by bee algorithm and response surface methodology. *J. Mol. Liq.* **2015**, *206*, 195–206. [CrossRef]

86. Yang, X.S. Firefly algorithm, stochastic test functions and design optimisation. *arXiv* **2010**, arXiv:1003.1409. [CrossRef]

87. Rashedi, E.; Nezamabadi-Pour, H.; Saryazdi, S. GSA: A gravitational search algorithm. *Inf. Sci.* **2009**, *179*, 2232–2248. [CrossRef]

88. Yazdani, S.; Nezamabadi-pour, H.; Kamyab, S. A gravitational search algorithm for multimodal optimization. *Swarm Evol. Comput.* **2014**, *14*, 1–14. [CrossRef]

89. Tharwat, A.; Hassanien, A.E. Quantum-Behaved Particle Swarm Optimization for Parameter Optimization of Support Vector Machine. *J. Classif.* **2019**, 1–23. [CrossRef]

90. Bandyopadhyay, S.; Saha, S.; Maulik, U.; Deb, K. A simulated annealing-based multi-objective optimization algorithm: AMOSA. *IEEE Trans. Evol. Comput.* **2008**, *12*, 269–283. [CrossRef]

91. Acharya, S.; Saha, S.; Sahoo, P. Bi-clustering of microarray data using a symmetry-based multi-objective optimization framework. *Soft Comput.* **2018**, 1–22. [CrossRef]

92. Bäck, T.; Rudolph, G.; Schwefel, H.P. Evolutionary programming and evolution strategies: Similarities and differences. In Proceedings of the Second Annual Conference on Evolutionary Programming, Los Altos, CA, USA, 25–26 February 1993.

93. Ferreira, C. Gene expression programming: A new adaptive algorithm for solving problems. *arXiv* **2001**, arXiv:cs/0102027.

94. Guven, A.; Aytek, A. New approach for stage–discharge relationship: Gene-expression programming. *J. Hydrol. Eng.* **2009**, *14*, 812–820. [CrossRef]

95. Koza, J.R.; Koza, J.R. *Genetic Programming: On the Programming of computers by Means of Natural Selection*; MIT Press: Cambridge, MA, USA, 1992.

96. Mitra, A.P.; Almal, A.A.; George, B.; Fry, D.W.; Lenehan, P.F.; Pagliarulo, V.; Cote, R.J.; Datar, R.H.; Worzel, W.P. The use of genetic programming in the analysis of quantitative gene expression profiles for identification of nodal status in bladder cancer. *BMC Cancer* **2006**, *6*, 159. [CrossRef] [PubMed]

97. Cheng, R.; Gen, M. Parallel machine scheduling problems using memetic algorithms. In Proceedings of the 1996 IEEE International Conference on Systems, Man and Cybernetics. Information Intelligence and Systems (Cat. No. 96CH35929), Beijing, China, 14–17 October 1996; Volume 4, pp. 2665–2670.

98. Knowles, J.D.; Corne, D.W. M-PAES: A memetic algorithm for multi-objective optimization. In Proceedings of the 2000 Congress on Evolutionary Computation. CEC00 (Cat. No. 00TH8512), Istanbul, Turkey, 5–9 June 2000; Volume 1, pp. 325–332.

99. Duval, B.; Hao, J.K.; Hernandez, J.C. A memetic algorithm for gene selection and molecular classification of cancer. In Proceedings of the 11th Annual Conference on Genetic and Evolutionary Computation, Montreal, QC, Canada, 8–12 July 2009; pp. 201–208.

100. Chehouri, A.; Younes, R.; Khoder, J.; Perron, J.; Ilinca, A. A selection process for genetic algorithm using clustering analysis. *Algorithms* **2017**, *10*, 123. [CrossRef]

101. Srivastava, A.; Chakrabarti, S.; Das, S.; Ghosh, S.; Jayaraman, V.K. Hybrid firefly based simultaneous gene selection and cancer classification using support vector machines and random forests. In Proceedings of the Seventh International Conference on Bio-Inspired Computing: Theories and Applications (BIC-TA 2012), Gwalior, India, 14–16 December 2012; pp. 485–494.

102. Babu, G.P.; Murty, M.N. Clustering with evolution strategies. *Pattern Recognit.* **1994**, *27*, 321–329. [CrossRef]

103. Bäck, T.; Schwefel, H.P. An overview of evolutionary algorithms for parameter optimization. *Evol. Comput.* **1993**, *1*, 1–23. [CrossRef]

104. Bäck, T.; Fogel, D.B.; Michalewicz, Z. (Eds.) *Evolutionary Computation 1: Basic Algorithms and Operators*; CRC Press: Boca Raton, FL, USA, 2018.

105. Eiben, A.E.; Smith, J. From evolutionary computation to the evolution of things. *Nature* **2015**, *521*, 476.

106. Lynn, N.; Ali, M.Z.; Suganthan, P.N. Population topologies for particle swarm optimization and differential evolution. *Swarm Evol. Comput.* **2018**, *39*, 24–35. [CrossRef]

107. Liu, Y.; Li, Z.; Xiong, H.; Gao, X.; Wu, J.; Wu, S. Understanding and enhancement of internal clustering validation measures. *IEEE Trans. Cybern.* **2013**, *43*, 982–994.

108. Karo, I.M.K.; MaulanaAdhinugraha, K.; Huda, A.F. A cluster validity for spatial clustering based on davies bouldin index and Polygon Dissimilarity function. In Proceedings of the Second International Conference on Informatics and Computing (ICIC), Jayapura, Indonesia, 1–3 November 2017; pp. 1–6.

109. Nies, H.W.; Daud, K.M.; Remli, M.A.; Mohamad, M.S.; Deris, S.; Omatu, S.; Kasim, S.; Sulong, G. Classification of Colorectal Cancer Using Clustering and Feature Selection Approaches. In Proceedings of the International Conference on Practical Applications of Computational Biology & Bioinformatics, Porto, Portugal, 21–23 June 2017; pp. 58–65.

110. Billmann, M.; Chaudhary, V.; ElMaghraby, M.F.; Fischer, B.; Boutros, M. Widespread Rewiring of Genetic Networks upon Cancer Signaling Pathway Activation. *Cell Syst.* **2018**, *6*, 52–64. [CrossRef] [PubMed]

111. Labed, K.; Fizazi, H.; Mahi, H.; Galvan, I.M. A Comparative Study of Classical Clustering Method and Cuckoo Search Approach for Satellite Image Clustering: Application to Water Body Extraction. *Appl. Artif. Intell.* **2018**, *32*, 96–118. [CrossRef]

112. Aarthi, P. Improving Class Separability for Microarray datasets using Genetic Algorithm with KLD Measure. *Int. J. Eng. Sci. Innov. Technol.* **2014**, *3*, 514–521.

113. Gomez-Pilar, J.; Poza, J.; Bachiller, A.; Gómez, C.; Núñez, P.; Lubeiro, A.; Molina, V.; Hornero, R. Quantification of graph complexity based on the edge weight distribution balance: Application to brain networks. *Int. J. Neural Syst.* **2018**, *28*, 1750032. [CrossRef] [PubMed]

114. Oyelade, J.; Isewon, I.; Oladipupo, F.; Aromolaran, O.; Uwoghiren, E.; Ameh, F.; Achas, M.; Adebiyi, E. Clustering algorithms: Their application to gene expression data. *Bioinform. Biol. Insights* **2016**, *10*. [CrossRef] [PubMed]

115. Tang, H.; Zeng, T.; Chen, L. High-order correlation integration for single-cell or bulk RNA-seq data analysis. *Front. Genet.* **2019**, *10*, 371. [CrossRef]

116. Kiselev, V.Y.; Andrews, T.S.; Hemberg, M. Challenges in unsupervised clustering of single-cell RNA-seq data. *Nat. Rev. Genet.* **2019**, *20*, 273–282. [CrossRef]

117. Handhayani, T.; Hiryanto, L. Intelligent kernel k-means for clustering gene expression. *Procedia Comput. Sci.* **2015**, *59*, 171–177. [CrossRef]

118. Shanmugam, C.; Sekaran, E.C. IRT image segmentation and enhancement using FCM-MALO approach. *Infrared Phys. Technol.* **2019**, *97*, 187–196. [CrossRef]

119. Masciari, E.; Mazzeo, G.M.; Zaniolo, C. Analysing microarray expression data through effective clustering. *Inf. Sci.* **2014**, *262*, 32–45. [CrossRef]

120. Bouguettaya, A.; Yu, Q.; Liu, X.; Zhou, X.; Song, A. Efficient agglomerative hierarchical clustering. *Expert Syst. Appl.* **2015**, *42*, 2785–2797. [CrossRef]

121. Lin, C.R.; Chen, M.S. Combining partitional and hierarchical algorithms for robust and efficient data clustering with cohesion self-merging. *IEEE Trans. Knowl. Data Eng.* **2005**, *17*, 145–159.

122. Darong, H.; Peng, W. Grid-based DBSCAN algorithm with referential parameters. *Phys. Procedia* **2012**, *24*, 1166–1170. [CrossRef]

123. Langohr, L.; Toivonen, H. Finding representative nodes in probabilistic graphs. In *Bisociative Knowledge Discovery*; Springer: Berlin/Heidelberg, Germany, 2012; pp. 218–229.

124. Carneiro, M.G.; Cheng, R.; Zhao, L.; Jin, Y. Particle swarm optimization for network-based data classification. *Neural Netw.* **2019**, *110*, 243–255. [CrossRef]

125. Yi, G.; Sze, S.H.; Thon, M.R. Identifying clusters of functionally related genes in genomes. *Bioinformatics* **2007**, *23*, 1053–1060. [CrossRef]

126. Somintara, S.; Leardkamolkarn, V.; Suttiarporn, P.; Mahatheeranont, S. Anti-tumor and immune enhancing activities of rice bran gramisterol on acute myelogenous leukemia. *PLoS ONE* **2016**, *11*, e0146869. [CrossRef]

127. Chavali, A.K.; Rhee, S.Y. Bioinformatics tools for the identification of gene clusters that biosynthesize specialized metabolites. *Brief. Bioinform.* **2017**, *19*, 1022–1034. [CrossRef]

 *processes*

Article

A Genetic Programming Strategy to Induce Logical Rules for Clinical Data Analysis

José A. Castellanos-Garzón [1,2,*], Yeray Mezquita Martín [1], José Luis Jaimes Sánchez [3], Santiago Manuel López García [3] and Ernesto Costa [2]

[1]   Department of Computer Science and Automatic, Faculty of Sciences, BISITE Research Group, University of Salamanca, Plaza de los Caídos, s/n, 37008 Salamanca, Spain; yeraymm@usal.es
[2]   CISUC, Department of Computer Engineering, ECOS Research Group, University of Coimbra, Pólo II - Pinhal de Marrocos, 3030-290 Coimbra, Portugal; ernesto@dei.uc.pt
[3]   Instituto Universitario de Estudios de la Ciencia y la Tecnología, University of Salamanca, 37008 Salamanca, Spain; a108985@usal.es (J.L.J.S.); slopez@usal.es (S.M.L.G.)
*   Correspondence: jantonio@usal.es

Received: 31 October 2020; Accepted: 26 November 2020; Published: 27 November 2020

Abstract: This paper proposes a machine learning approach dealing with genetic programming to build classifiers through logical rule induction. In this context, we define and test a set of mutation operators across from different clinical datasets to improve the performance of the proposal for each dataset. The use of genetic programming for rule induction has generated interesting results in machine learning problems. Hence, genetic programming represents a flexible and powerful evolutionary technique for automatic generation of classifiers. Since logical rules disclose knowledge from the analyzed data, we use such knowledge to interpret the results and filter the most important features from clinical data as a process of knowledge discovery. The ultimate goal of this proposal is to provide the experts in the data domain with prior knowledge (as a guide) about the structure of the data and the rules found for each class, especially to track dichotomies and inequality. The results reached by our proposal on the involved datasets have been very promising when used in classification tasks and compared with other methods.

Keywords: clinical data; feature selection; genetic programming; machine learning; data mining; evolutionary computation

---

1. Introduction

Current data management and storage methods have been challenged by the high increase in the amount of medical data available to us. Obtaining valuable information in the process of knowledge discovery has become problematic. There is an urgent need for new tools and approaches whose mechanism will allow overcoming the present-day limitations of computational medicine, by converting large quantities of data into knowledge. Novel methods will make it possible to go beyond simple data description, providing knowledge in the form of models. Through abstract data models, it is possible to create highly reliable prediction systems [1–14].

The process of knowledge discovery from the data involves, among other techniques, machine learning. Our interest is to select or combine techniques with a high performance in prediction tasks for medical datasets. In medicine, prediction systems are most frequently applied in the field of diagnosis and prognosis. According to previous research on the development of diagnosis systems, it is possible to determine the presence or absence of a disorder through interpretation of patient data [15]. These systems

Processes **2020**, *8*, 1565; doi:10.3390/pr8121565　　　　　　　　www.mdpi.com/journal/processes

are used specifically in the diagnosis of patients. Prognosis systems use the collected information to predict the progress of the condition a patient is suffering from or to determine whether a patient may suffer a disease in the future. Moreover, they are used to choose the most effective treatment based on the patient's symptoms and different medical factors [16].

In the context of diagnosis and prognosis, the aim of using intelligent systems based on machine learning techniques is knowledge discovery from the collected information. Sometimes, the discovered knowledge is expressed in a probabilistic model by relating the clinical features of patients to a stage of the target disease. In other cases, a rule-based representation is selected to provide the expert with an explanation of why certain decision was made. Knowledge representations as those described above are known as white box systems and the focus of this research because they express part of the knowledge directly. Finally, there are other cases in which the system is designed as a black box for decision-making, where the system only shows the prediction results. All of these techniques are suitable for making a diagnosis and prognosis of a patient's condition [17].

Because of all previously explained, this research proposes a system generating classifiers based on genetic programming (GP), which is capable of inducing sets of rules that represent the relationship between the disease and the symptoms experienced by patients. Therefore, our goal is to build a rule-based classifier and compare its ability to correctly classify data with other previously proposed methods. Finally, we analyze the rules obtained by our approach to determine the most important attributes of the dataset. In this case, the system performs a feature filtering process [18–21]. Rule-based classifiers are an attractive approach since the structure of IF/THEN rules is well-known and can easily be interpreted for knowledge discovery. Hence, such rules not only classify unknown patterns, they also disclose knowledge about the class structure and problem domain. The goal of a rule-based classifier is to find a set of rules that suit a labeled dataset. That is, the discovered rules should represent the target dataset and cover each region of the search space. Hence, the application of GP in the building of rule-based classifiers has been the basis of works such as [22–25]. Our ultimate goal is to provide the expert with an initial interpretation of the data through our rules-based model that can serve as a starting point in the study of the disease. Hence, we also provide a visual interpretation of the data, which supports the process of knowledge discovery.

In summary, medical databases store a lot of data about the health condition of patients. Such an amount of information is ideal for the application of machine learning techniques, which can transform data into knowledge by analyzing the relationships provided by the model. This mechanism provides a means of hypothesis validation [6,9]. To reach the goals proposed in this work, the rest of this manuscript has been divided into the following sections: Section 2 deals with the background related to this research. Section 3.1 describes the main features of our proposal, encoding, fitness functions, genetic operators and running strategy. Section 4 describes the employed datasets, an analysis of the structure and distribution of the datasets, the experiments to select the best mutation operators for each medical dataset, and accuracy comparison of our approach with other machine learning methods. At the end of this section, an analysis of the rules discovered by the proposal is given and the most influential attributes of the datasets are analyzed. Conclusions, Appendix A (classifiers of our proposal), Appendix B (mutation operator experiments), and the references of this research are the final part of this document.

2. Background

Applications of genetic algorithms (GAs) to analyze medical data have allowed for solving complex problems such as disease screening, diagnosis, treatment planning, pharmacovigilance, prognosis and health care management [26]. GAs have been applied to different fields in medicine, among which we can highlight, Radiology, Oncology, Cardiology, Endocrinology, Pulmonology and Pediatrics, among others. In this context, GAs have been used for edge detection of images obtained from Magnetic Resonance Imaging

(MRI), Compute Tomography (CT) and ultrasound [27–29]. Making use of these kinds of algorithms, different methods have been proposed to detect microcalcifications in mammograms leading to diagnosing breast cancer [30–32]. In other studies, GAs have been used to fuse MRI images with Positron Emission Tomography (PET) in order to generate colored images of breast cancer [33].

In other works [34], a methodology based on the application of a Micro-Genetic Algorithm (MGA) was used to generate the training set that best detects solitary lung nodules. The designed algorithm can detect lung nodules with about 86% sensitivity, 98% specificity, and 97.5% accuracy. In [35], the authors proposed a model using Particle Swarm Optimization method (PSO), a GA and a Support Vector Machine (SVM) in conjunction with feature selection and classification of CT, MRI and ultrasound images. The proposed method was capable of detecting lung cancer with an accuracy of 89.5%.

GAs have also been used to detect patients with some type of carcinoma through Microarray Technology. For example, in [36], a GA combined with an Artificial Bee Colony (ABC) algorithm was proposed. The method aims to make cancer classification in patients through extraction of features from microarray data. This method was tested with a dataset of colon carcinoma, two different datasets of Leukemia, a dataset involving patients with lung carcinoma, and one of patients with Small, Round-Blue Cell Tumors (SRBCT). The method proposed in that paper achieved an accuracy of almost 100% when selecting very few biomarkers.

In the area of Pediatrics, GAs are also being used to detect diseases such as autism from gene expression microarrays. In [37], an approach of GA as a feature selection engine and an SVM as the classifier were proposed to validate the set of features selected. In this work, a performance greater than 86% accuracy for one of the used datasets and a performance of 92.93% accuracy for the other dataset were reached to outperform previous works.

There are other applications of GAs aimed at making predictions from the data acquired from blood tests. In [38], a GA is used to optimize the performance of an Artificial Neural Network (ANN) to detect Coronary Artery Disease (CAD). Through the previous approach, the authors show that CAD can be detected without angiography and consequently eliminate its high cost and the main side effects. In another context, electrocardiogram (ECG) signals in cardiology have been used to detect cardiac arrhythmias [39]. In this work, a method liking a Genetic Algorithm with a Backpropagation Neural Network (GA-BPNN) was proposed to reduce the dimension of the datasets by 50% and achieve 99% accuracy. This makes the method suitable for automatic identification of cardiac arrhythmias.

As stated at the beginning of this section, there are many more applications of GAs to medicine that can be consulted about in the literature [40–42]. Since the efficacy of GAs in the medicine field has been proved, we will deal with other recent algorithms (Genetic Programming), which include a GA as its base operation. Genetic Programming (GP) is a kind of GA whose main difference with respect to normal GAs is to produce expressions (functions or programs) as outputs rather than data [43–45]. An example of the use of this kind of algorithms in the medical field is shown in [46]. In this work, a GP algorithm is proposed to automatically create the best mathematical formula that combines a set of preselected features from a Magnetoencephalography (MEG) dataset. To evaluate the generated formulas, a K-nearest neighbor algorithm (KNN) is used. This approach achieved 91.75% sensitivity and 92.99% specificity in the diagnosis of Epilepsy.

GP is also used to provide diagnosis from MRI images by evaluating the medical spine condition of patients [47]. The GP algorithm proposed in this work uses of a fitness function based on expert knowledge, in this case, a neuroradiologist. The rules rendered in each generation of the algorithm are evaluated and then compared with the true results in order to select the rules with less difference. The accuracy reached was greater than 90% in the conditions evaluated by combining the GP algorithm and expert knowledge.

Another example of GP applied to medicine is image classification [48]. In this work, a GP algorithm is proposed to create and evolve tree-based classifiers, whose aim is to diagnose active tuberculosis from raw X-ray images. The framework proposed was able to achieve a competitive classification and a superior speed compared to methods that rely upon image processing and feature extraction.

In general terms, GP represents a flexible and powerful evolutionary technique that uses a set of functions and terminals to produce computable expressions. Hence, this research presents a GP method to render rule-based classifiers for knowledge discovery from medical data. Some of the advantages related to this kind of classifiers generating comprehensible knowledge are high expressiveness, which allows them to render models that are very easy to interpret. Such rules can be altered to handle missing values and noise from attributes of the data set. They are relatively easy to obtain and very fast at classifying new patterns (or data) [49]. Moreover, a very important advantage of such rules for machine learning is that they are intuitively comprehensible to the user [50,51]. Another advantage related to the above is that they are not only used to classify, but they also represent, by themselves, a process of knowledge discovery, providing the user with new insights into the data and their application domain [52].

3. Materials and Methods

3.1. Evolutionary Strategy to Build Rule-Based Classifiers (ESRBC)

This section presents our main proposal, the evolutionary method (ESRBC) to render rule-based classifiers. Thus, we describe the strategy to follow by ESRBC, individuals, crossover, mutation operators and fitness functions. Individuals represent logical rules adopting an internal representation of a linear sequence of clauses (or comparisons) separated by conjunctions AND. Individuals to be built in this proposal follow the Michigan-style [24,50,53,54]; hence, each individual encodes a single rule (with a linear chromosome) with a variable length, where each rule is associated with the class of the dataset it represents. Therefore, an individual can be evaluated as True or False according to the pattern evaluated in the antecedent of the rule. As applicable, the pattern may or may not belong to the class assigned to the rule.

As explained, the individuals generated by ESRBC represent logical rules of type *IF <CLAUSES> THEN <CLASS>*, where *<CLAUSES>* is formed by a set of clauses (or comparisons) separated by conjunctions AND. *<CLASS>* is the class of the dataset that is being represented by the rule or, in other words, the class to which the rule belongs. A more detailed representation of a rule can be given as follows:

$$IF\ (at_1\ o_1\ val_1)\ AND\ (at_2\ o_2\ val_2)\ AND\ \cdots\ AND\ (at_n\ o_n\ val_n)\ THEN\ class = k,$$

where $(at_i\ o_i\ val_i)$ is clause number i, at_i is an attribute of the dataset, o_i is a comparison operator from set $\{<,>,\leq,\geq,=,\neq\}$, val_i a value of the set of all possible values admitted by at_i, whereas k is the class of the dataset covered by the rule. An example of logical rules representing a dataset with attributes $\{p,q,r,s,t\}$ and two classes $\{0,1\}$ can be as follows:

$$IF(p > 12.3)\ AND\ (p \leq 15)\ AND\ (s \neq 3.4)\ THEN\ class = 0,$$

$$IF(p \leq 12.3)\ AND\ (r > 7.4)\ AND\ (t \geq 2)\ THEN\ class = 1,$$

which means that, if there is a specific pattern $(p_i, q_i, r_i, s_i, t_i)$ from the domain of the dataset, whose values p_i and s_i hold the antecedent of the rule in class-0, then such a pattern belongs to class-0. Likewise, if attribute values p_i, r_i, t_i hold the antecedent of the rule in class-1, then this pattern is in class-1. Keep in mind that the challenge that each rule learned from a dataset must meet is generalization. In other words, the set of rules holding a dataset should generalize enough in such a way that the pattern space be properly partitioned. Thereby, each region of the space is covered as much as possible by the set of rules.

Continuing with the description of the rule concept, we define the length of a rule as the number of clauses that form it. The evolutionary algorithm (EA) of our approach, which is responsible for the search process for a diverse set of rules, adopts the sequential covering strategy for each class of a dataset [51]. Sequential covering is a technique that discovers one rule at a time. The EA is executed multiple times to build a complete set of rules representing each class of a dataset. During each execution, the best rule evolved through the EA is added to the set of previously discovered rules and the patterns covered by this rule are removed from the dataset. The process is repeated until there are no more patterns to be covered. The steps followed by this methodology can be described as follows:

1. Select the set of patterns from a new class i in the input dataset;
2. Create an initial population P_0 of rules candidate to represent patterns in class i;
3. Run the EA on P_0 to achieve a final population P_f;
4. Add the most fit individual (rule) r of P_f to the set of rules R (R is empty initially);
5. Remove all patterns from class i holding rule r;
6. If class i is not empty, then go to step 2;
7. If there are more classes in the dataset, then $i := i + 1$ and go to step 1;
8. At the end of the process, R has a set of rules learned from each class of the input dataset.

3.2. Fitness Functions

This section introduces the fitness functions used in the evolutionary algorithm of our approach. In this case, the fitness functions defined are based on the concept of accuracy [52,55,56]. The accuracy of a rule is the fraction of patterns from its class, covered by the rule. Then, according to the definition above, we are going to introduce two variants of fitness functions based on accuracy. However, we firstly need to define two functions which evaluate a pattern e in a rule r. Then, the first function is g acting on r and e, i.e., $g(r,e)$, which computes the number of clauses of r evaluated True when e is evaluated in r. The second function defines the evaluation of a pattern e in r ($r(e)$) in the following way:

$$r(e) = \begin{cases} 1, \text{if } e \text{ belongs to the class of } r, \text{ in this case we say, } e \text{ holds } r; \\ 0, \text{otherwise.} \end{cases} \tag{1}$$

Note that $g(r,e)$ evaluates the number of clauses in r holding a pattern e, whereas $r(e)$ evaluates the rule to 1 (True) if it covers pattern e (all its clauses become True). Additionally, if we want to specify the class of both r and e, we write r^i and e^i respectively, where i is a class of the dataset. Finally, the two expected fitness functions are given below. For this case, both fitness functions define a maximization problem. The first objective of f_1 assesses accuracy based on the number of clauses turned true by patterns of the target class, whereas the second objective acts as a penalty for patterns not belonging to the class of the rule, whose values make the clauses of the rule true. The same situation happens for f_2, but, in this case, the accuracy is assessed by considering the number of patterns holding a rule r. f_1 has been created to be run in the first generations of the evolutionary algorithm where rules have randomly been created and no pattern holds them. However, the use of f_2 makes more sense in a second stage of the evolutionary algorithm (after applying f_1) when the rendered rules have reached a certain learning level.

Definition 1. *Fitness function-f_1.*
If D is a labeled dataset with k classes, C_i a class of D and r^i a rule of C_i and consider $i, j \in [0, 1, \cdots, k-1]$. Then, we define a fitness function-f_1 applied to r^i as:

$$f_1(r^i) = \frac{1}{|C_i| \cdot |r^i|} \sum_{\forall e \in C_i} g(r^i, e) - \frac{1}{|D| - |C_i|} \sum_{\forall e' \in C_j, j \neq i} g(r^i, e') + 3. \tag{2}$$

Definition 2. *Fitness function-f_2.*
From the same conditions given in Definition 1, we define fitness function-f_2 applied to a rule r^i as:

$$f_2(r^i) = \frac{1}{|C_i|} \sum_{\forall e \in C_i} r^i(e) - \frac{1}{|D| - |C_i|} \sum_{\forall e' \in C_j, j \neq i} r^i(e') + 3. \tag{3}$$

Both fitness functions have been focused on a maximum problem: the bigger their values, the more fit the evaluated rules. In the first fitness function, the first objective deals with a kind of accuracy using g, which consists of computing the number of clauses evaluated True in the current rule for all pattern of its class. The second objective measures the number of clauses evaluated True by the current rule for all pattern belonging a different class of the rule class. This fitness function is useful in the evaluation of rules built in the first generations of the EA, where the rule accuracy is zero. The second fitness function is responsible for measuring the number of patterns from the rule class holding the rule versus the number of patterns of other classes holding the rule.

3.3. Genetic Operators

The crossover operator used in this method to recombine clauses from two parent-rules to achieve two new children-rules performs as the classical operator [57]. That is, the crossover operator selects a random position (with a uniform distribution) from two parent-rules and exchanges two segments of clauses from them to achieve two children, inheriting part of the clauses (genetic code) of their parents. In other words, given two rules, the position of a clause is randomly selected. Then, the clauses located on the right or left side of both rules (which is also decided at random) are exchanged to create two new rules. The mutation operator is responsible for providing new information to the individuals generated. In this case, we provide three types of mutation operations by defining a mutation group for each one:

1. *Mutation by clause, M1*: Changes the attribute, comparison operator or value in a randomly chosen clause from the rule by others, also randomly selected;
2. *Mutation clause by clause, M2*: This operator applies the M1 operator clause by clause to a rule. For each clause, the operator decides whether to mutate. If a mutation has been selected, then the operator decides what mutation type to perform. Namely, changing the attribute, the comparison operator or the value of the attribute in the current clause.
3. *Mutation by transformation, M3*: This operator can remove a part of the rule, add a new rule, or apply the M1 operator to the rule. One of the three operations above is selected at random. In the first operation, a position in the rule is randomly selected to remove the left or right side. Then, it randomly selects the part of the rule to be removed. The second operation adds a new rule at the end of the current rule. The added rule is randomly created (by also choosing its size in a random way).

The mutation operator applied to each mutation in the rules is selected at random. Note also that the goal of defining the M2 and M3 compound mutation operators is to create different mutation levels from the M1 basic mutation operator. This allows us to explore different alterations on the individuals yielded from generation to generation. Each of these operators (M1, M2, and M3) performs an alteration level of individuals by regarding a minor (M1), medium (M2) and higher level (M3) of alteration.

3.4. Running the Evolutionary Algorithm

Once the genetic operators have been defined, the evolutionary algorithm (EA) of our proposal ESRBC is responsible for discovering each rule covering different parts of the search space, hoping the rules can generalize. Hence, the EA is run following the general scheme given by evolutionary algorithms [57,58], with the particularity of introducing an elitism which is transmitted from generation to generation and tournament selection as the adopted selection method.

Aside from the above, the EA includes an evolutionary strategy of local search (Algorithm 1 ESLS), which acts on the population or the most fit individual returned by the EA. In fact, the option of executing Algorithm 1 from a population or a single individual is a parameter of the algorithm. The term local search is because Algorithm 1 is based on mutation operators and in each generation, Algorithm 1 replaces only individuals who have improved their value fitness after the mating process. The goal of this strategy is to refine the solutions of the EA by making an in-depth search. Hopefully, the individuals from the EA are close enough to a global optimum. Therefore, Algorithm 1 is in charge of searching such an optimum. This idea has been taken from [59] and implemented in [60] with good results. The idea is as follows:

- *Running a genetic algorithm (GA) until it slows down, then letting a local optimizer take over the last generation (and/or best individual) of the GA. Hopefully, the GA is very close to the global optimal.*

ESLS has been defined below. This strategy improves a population of individuals or a single individual given by ESRBC. Finally, both ESRBC and Algorithm 1 were implemented in the *C++ programming language*, whereas the experiments were performed under *R-Project* [61].

Algorithm 1 ESLS

Input: POP, the population composed by the individuals in the last generation of the EA. MaxGeneration, the number of generations. MO applies one of the mutation operators given in Section 3.3, chosen at random. f_2, fitness function given in Section 3.2.

Output: POP, as a result of improvement of the input.

---

1. $t := 0$;
2. **while** $t <$ MaxGeneration **do**
3. $t := t + 1$;
4.   **for all** rule r in POP **do**
5.   % Computing fitness
6.   f := $f_2(r)$;
7.   % Applying mutation.
8.   newr := MO(r);
9.   % Evaluating new individual.
10.   newf := f_2(newr);
11.   % Updating the improved individual.
12.   **if** newf > f **then** $r :=$ newr;
13.   **end for**
14. **end while**
15. **end.**

---

4. Results

This section describes the experiments carried out by our proposal on the clinical datasets, which have been selected from Machine Learning Repository [62]. The used datasets deal with Heart, Hepatitis, and Dermatology diseases, which we call DS1, DS2, and DS3, respectively, and have the features listed below. Note that Number of patterns refers to the number of instances (number of rows) of the dataset while Number of attributes refers to the number of variables (number of columns) of the dataset.

1. Title: SPECTF Heart Dataset (DS1);

    - Number of patterns: 267;
    - Number of attributes: 22 plus the class attribute;
    - Number of classes: 2 classes;
    - Attribute type: binary;
    - Missing attribute values: No missing values.

2. Title: Hepatitis Domain (DS2);

    - Number of patterns: 155;
    - Number of attributes: 19 plus the class attribute;
    - Number of classes: 2 classes;
    - Attribute type: Categorical, Integer and Real;
    - Missing attribute values: yes (10-nearest neighbor technique was used in this research for imputation of missing values).

3. Title: Dermatology Database (DS3);

    - Number of patterns: 366;
    - Number of attributes: 34 plus the class attribute;
    - Number of classes: 6 classes;
    - Attribute type: Categorical and Integer;
    - Missing attribute values: yes (20-nearest neighbor technique was used in this research for imputation of missing values).

4.1. Exploring and Analyzing the Datasets

This section shows different linked views of the distribution and structure of the datasets, which allow us to have an initial assessment of their behavior. This can also help explain some of the results obtained for the methods applied. Starting with the DS1 dataset shown in Figure 1, we have that this dataset is represented by a diamond-shaped cloud of points. According to the point distribution in each class, Class-1 is much more compact and bigger than Class-0. Therefore, Class-0 could need more rules to classify the patterns of its class than Class-1. Since points in Class-1 are more scattered in space and both classes are intertwined, it would be more difficult for this class to find rules that do not classify patterns in Class-0 by mistake. On the other hand, at the bottom of the figure, there is the heatmap of the dataset where both classes are separated by boxes. Note that the values in this dataset are binary and, in Class-0, values 0 predominate while, in the other class, values 1 predominate. This means that, unlike Class-1, Class-0 is characterized by the absence of the property denoted by many of the attributes evaluated by the disease represented in the dataset.

Turning now to the DS2 dataset shown in Figure 2, we have that this dataset is represented by a cloud of points in tree form at the top of the figure. As in the DS1 dataset, both classes are intertwined, Class-1 is more compact and bigger than Class-0. Unlike the DS1 dataset, points are more scattered in space, which can induce a smaller number of rules classifying each class. Note that it is more difficult to find visual differences separating both classes for the heatmap given for DS2 than for the case of DS1. The above can imply that DS2 is a difficult dataset to classify, which tests any applied classifier (method).

As for the DS3 dataset shown in Figure 3, we have that, unlike the other datasets, this one has six classes, which may increase the classification error of the methods applied to the dataset. The point cloud of this dataset, shown at the top of the figure, is T-shaped with agglomerations of points at the ends and in the center of the 3D-scatterplot. Note that the same T-structure of the dataset is maintained for the points in each class. In this case, each class may generate four rules since there are four clusters in each class. However, the greatest difficulty would be to separate the classes from the others, since the six classes are very interrelated. Finally, note that classes 0, 1, and 2 differ from classes 3, 4, and 5 in that, in the latter, the light green color predominates (values below the average value of the whole dataset), while, in the rest of classes, the representative color is brown (values above the average value of the whole dataset). This shows that classes 0, 1, and 2 share some type of similarity with the type of disease represented by each class, which makes the difference from the diseases represented in classes 3, 4, and 5. The same reasoning done for classes 0, 1, and 2 is met for classes 3, 4, and 5 of DS3.

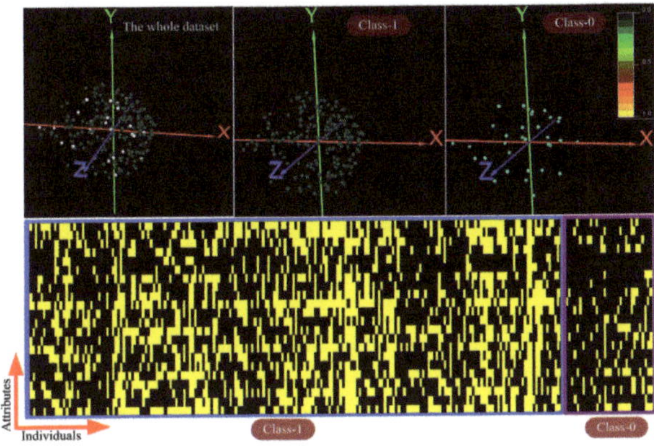

Figure 1. DS1 dataset (Heart Dataset). A 3D-scatterplot is shown at the top of the figure where each point represents a column (an individual) of the dataset showed as a heatmap at the bottom. The dimension of the dataset was reduced to three components by using principal component analysis. In addition, points belonging to each class are shown in different colors. The heatmap corresponding to the same dataset is shown at the bottom of the figure. Each class of the dataset is framed in a box. The color bar shown at the top represents the color scale used in the heatmap.

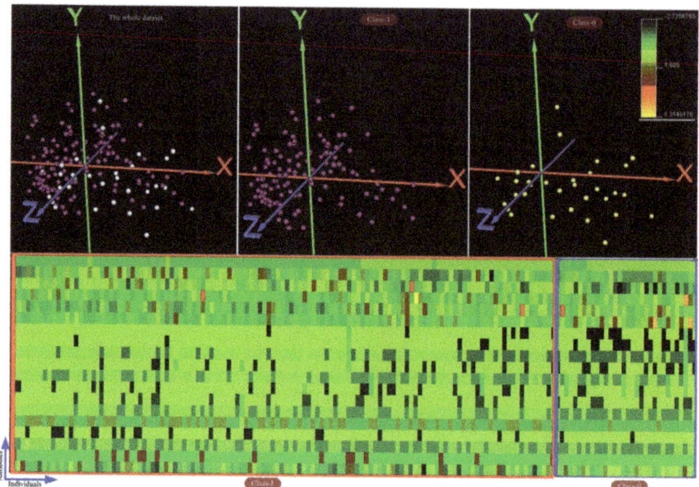

Figure 2. DS2 dataset (Hepatitis Dataset). A 3D-scatterplot is shown at the top of the figure where each point represents a a column (an individual) of the dataset showed as a heatmap at the bottom. The dimension of the dataset was reduced to three components by using principal component analysis. In addition, points belonging to each class are shown in different colors. The heatmap corresponding to the same dataset is shown at the bottom of the figure. Each class of the dataset is framed in a box. The color bar shown at the top represents the color scale used in the heatmap.

4.2. Mutation Operator Evaluation

This section deals with the evaluation of mutation operators M1, M2, and M3 from their effectiveness and behavior under the different given datasets. The goal of this test is to carry out an analysis of the behavior of the mutation operators to select the operators having a better performance on a given dataset. Consequently, we have run the evolutionary method (ESRBC) using only the mutation operators (without the crossover operator). Then, ESRBC is run 20 times for each operator and each dataset (DS1, DS2, and DS3) in the following way: for each mutation operator applied to a dataset, ESRBC has been run 20 times, each using a different mutation probability value. In each execution, the probability value is increased in a step of 0.05, starting from 0. Then, for each mutation value, the fitness value of the most fit individual in 5000 generations has been taken out to render the graphics given in Figures A1–A3 (Appendix B) for datasets DS1, DS2, and DS3, respectively. Thus, the achieved graphics represent mutation probability values in x-axis versus fitness value (y-axis) for the best individual yielded in each mutation probability value.

As shown in these figures, each row deals with four graphics, which correspond to the same experiment repeated with the same mutation operator. Since ESRBC includes a stochastic process in the search, we have repeated the experiment four times for each mutation operator. Therefore, each row in these figures correspond to a mutation operator, i.e., the first row represents M1, the second and third rows correspond to M2 and M3, respectively. Finally, each graphic in each row represents the fitness values reached by the best individuals for the current operator mutation. Such values are represented by means of a blue curve. The mean fitness values from the four experiments carried out (in each row) for each operator are represented through the green curve, whereas the standard error bars are stressed in pink lines.

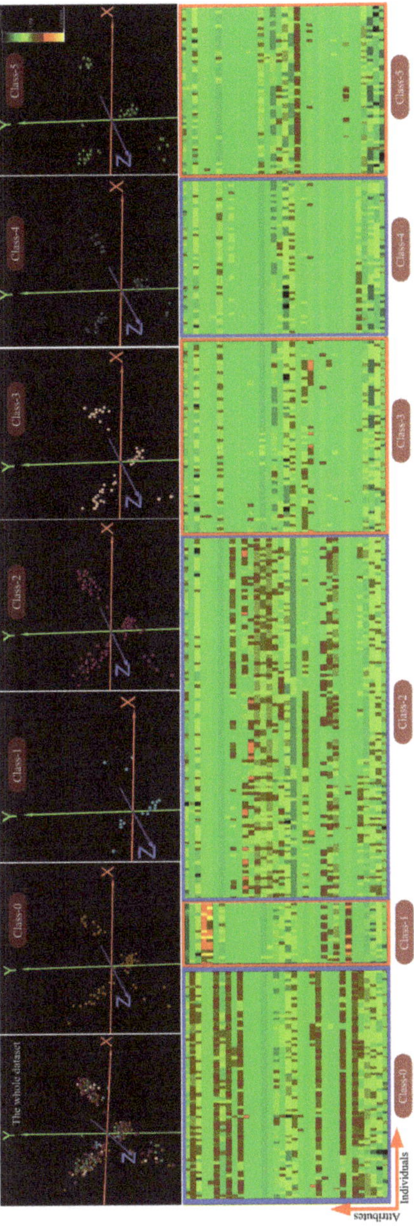

Figure 3. DS3 dataset (Dermatology Dataset). A 3D-scatterplot is shown at the top of the figure where each point represents a column (an individual) of the dataset showed as a heatmap at the bottom. The dimension of the dataset was reduced to three components by using principal component analysis. In addition, points belonging to each class are shown in different colors. The heatmap corresponding to the same dataset is shown at the bottom of the figure. Each class of the dataset is framed in a box. The color bar shown at the top represents the color scale used in the heatmap.

By making now an analysis of the results for these three figures, we can say that, for the DS1 dataset, the results in Figure A1 (Appendix B) show that, for operators M1 and M2, most of the reached fitness values (blue curve) are between 3 and 3.4. On the other hand, there is overlap between the error bars (pink bars), which indicates uniformity of fitness values for M1 and M2 with respect to different mutation probability values. However, the fitness values (blue curve) given in the graphics for operators M1 and M2 present more oscillations than the ones represented by the curves given for the M3 operator. In addition, the standard error bars for the M3 operator are smaller, indicating that the average value plotted is more reliable than the one of those in M1 and M2. Moreover, most fitness values with respect to mutation probability values are between 3.2 and 3.4. Thus, the M3 operator appears to be more significant for the DS1 dataset than operators M1 and M2. Hence, we can use only M3 as the mutation operator when using the evolutionary method to build a classifier on the DS1 dataset. In addition, keep in mind that, since the standard error bars overlap in the M3 operator, it is not necessary to assign a big valor of mutation probability in the running of ESRBC to build the rule-based classifier. The above improves the runtime of the method.

Unlike Figure A1 (Appendix B), graphics in Figure A2 present more oscillations according to the curve representing fitness values across from mutation probability values (blue curve). However, the error bars maintain an overlap. In addition, note that the fitness values achieved for M1 and M3 are higher than those given in Figure A1. That is, for M1, most fitness values are between 3.4 and 3.6. For M3, most fitness values are between 3.5 and 4 whereas fitness values for the M2 mutation operator are more unstable with respect to mutation probability values. Therefore, we can use operators M1 and M3 as the only ESRBC mutation operators when using the DS2 dataset.

For the results obtained from the DS3 dataset, Figure A3 (Appendix B), we have that they are like those given in Figure A1. Hence, by applying the same reasoning as the one given in Figure A1, the M3 mutation operator is the most stable and so the operator that best performs on the DS3 dataset. Once the mutation operators performing well on each dataset have been chosen, we can proceed to compare the classifiers induced by our method with other machine learning methods under the accuracy measure.

4.3. Accuracy and Comparison of the Evolutionary Method

The accuracy of the rule-based classifier yielded by our approach has been computed and compared with other methods for each introduced dataset. A stratified 10-fold cross-validation was used to measure the accuracy of all methods. The evolutionary method (ESRBC) defined was run in two stages. In the first stage, ESRBC was run by using the f_1 fitness function, whereas f_2 was used in second stage. The settings of ESRBC for each dataset have been listed in Table 1 and the methods used in the comparison process [55,56,63,64] have been listed in Table 2. Then, the results reached by ESRBC compared with other methods have been listed in Table 3. The best accuracy value for each dataset has been underlined. ESRBC reached the best values for the DS1 and DS3 datasets while its accuracy for DS2 was not very different from the one of the method reaching the best value. Since the number of patterns for the classes of the DS1 and DS2 datasets are unbalanced, Table 3 also shows the Youden index which deals with unbalanced classes in a dataset. This index is defined as $sensitivity + specificity - 1$.

Note that the methods listed for the DS3 dataset are different from those used in DS1 and DS2. This is because the methods used for DS1 and DS2 are for binary classification, whereas the DS3 dataset has six classes. Therefore, we need to use multiclass methods in DS3, different from the methods used in the previous datasets. On the other hand, the greatest accuracy reached for the DS1 and DS2 datasets was less than 90%, which tells us that they are difficult to classify (due to their compactness and difference in the size of their classes), as explained in Section 4.1. However, the greatest accuracy reached for the DS3 dataset was greater than 90%, although DS3 has six classes. This may be due to the distribution of the dataset, where each class is represented by four groups of points separated from each other (which facilitates the classification), as explained in Section 4.1.

Table 1. Settings given to run the evolutionary method (ESRBC) to build rule-based classifiers for each dataset.

ESRBC Settings	DS1	DS2	DS3
Population size	70	50	100
Number of generations per rule (fitness function-1)	100,000	100,000	200,000
Number of generations per rule (fitness function-2)	200,000	200,000	400,000
Mutation operator per rule for Algorithm 1 (local search)	10,000	10,000	100,000
Mutation operator	M3	M1 and M3	M3
Crossover probability	0.6	0.6	0.6
Mutation probability	0.2	0.3	0.2
Maximum size of rules (maximum number of clauses)	10	10	10

Table 2. Name and description of the methods used in the comparison of the approach proposed.

Method	Description
SVM	Linear Support Vector Machine, which finds the best hyperplane separating both classes.
naiveBayes	This model computes the probability of each class given the values of all attributes and assuming the attribute conditional independence.

Table 2. *Cont.*

Method	Description
kNN	k-Nearest Neighbor classification. This is a lazy model which classifies the input pattern by using its k-Nearest Neighbors from the training set.
ANN	Artificial Neural Network implementing a Multilayer Perceptron, which uses a single intermediate layer for our case. Backpropagation and resilient backpropagation have been implemented.
ML-MPCA	Maximum Likelihood estimation with Mixture of Principal Component Analyzers.
Bayesian-MPCA	Bayesian approach with Mixture of Principal Component Analyzers.

Table 3. Comparative table of mean accuracy for the evolutionary method (ESRBC) compared with the other machine learning methods.

Dataset	Method	Accuracy (%)	Youden Index (%)
DS1	SVM	78.95	12.45
	naiveBayes	46.93	14.97
	k-NN (k = 3)	81.58	22.91
	ANN	79.39	12.95
	ESRBC	<u>81.82</u>	<u>40.48</u>
DS2	SVM	<u>82.57</u>	<u>46.91</u>
	naiveBayes	75.47	37.56
	k-NN (k = 5)	75.47	16.47
	ANN	81.94	42.56
	ESRBC	81.32	41.67
DS3	ML-MPCA	94.9	-
	Bayesian-MPCA	95.8	-
	ESRBC	<u>95.92</u>	-

5. Discussion: Rule Analysis

This section makes an analysis of rules discovered by the classifiers induced by the evolutionary method (ESRBC) for each dataset in Table 3. The aim of this analysis is to discover knowledge from those rules and identify attributes and relations relevant for the disease. In that sense, such prior knowledge would act as a starting point for experts in this field.

Appendix A lists the rules given by the best classifier found by our proposal for each dataset. The analysis carried out in this section is based on knowledge disclosed from such rules. Starting from the DS1 dataset, we have that it contains diagnoses based on 22 features, built from Single Proton Emission Computed Tomography (SPECT) images, which aim to distinguish between heart disease and normal heart operation. For this case, ESRBC found six rules for a class and only one rule for the other class. Of the 22 attributes, only five of them were not used by any rule (F1, F2, F9, F12, and F18), which implies that they are not important in the classification of the disease and may be discarded from the analysis. However, attributes F5, F21, and F22 achieved the greatest frequency of occurrence by rules in class-0 (they occurred in 42.86% of rules). Hence, such attributes are representative for class-0. Meanwhile, class-1 used a single rule with only one attribute, F8. The F8 attribute has been used in both classes, so it is not only important for class-1 but also for the disease in question.

The DS2 dataset consists of 19 attributes and two different classes, including clinical and biochemical variables. ESRBC found three rules for class-0 and 2 rules for class-1. Of the 19 attributes, 12 of them were used in the rules and seven of them were not used by any rule (STEROID, MALAISE, ANOREXIA, LIVERBIG, LIVERFIRM, VARICES and HISTOLOGY). Thus, they can be discarded from the classification process. In addition, the most frequent attributes by rules in class-0 were ALBUMIN with 100%, PROTIME, AGE and ALK PHOSPHATE with 66.67%, whereas class-1 only used the ALBUMIN and SEX attributes. Note that the ALBUMIN attribute is the only one used in both classes. Therefore, this attribute is significant for the study of the disease. This way, we can identify three groups of patients presenting different features in class-0: patients holding $\{ALBUMIN \leq 3.99, PROTIME \leq 50, SEX = 1\}$, patients holding $\{ALBUMIN \neq 3.80, AGE \geq 37, 64.88 \leq ALKPHOSPHATE < 95\}$ and patients holding $\{ALBUMIN \geq 3.50, PROTIME \geq 56.16, ALKPHOSPHATE > 104.77, AGE \geq 30\}$. Patients in class-1 are government by attributes $\{ALBUMIN \geq 2.9, SEX = 2\}$.

The DS3 dataset presents data of patients of six different erythemato-squamous diseases. That is, *psoriasis, seboreic dermatitis, lichen planus, pityriasis rosea, cronic dermatitis* and *pityriasis rubra pilaris*. The main interest of applying our proposal to this dataset is that these diseases are difficult to distinguish, and they normally require a biopsy and present many common histologic characteristics. The classifier found for DS3 rendered 11 rules distributed in the six classes. Namely, 1 rule in classes 0, 2, 4 and 5; 2 rules in class-1 and 4 rules in class-3, which coincides with that explained in Section 4.1 for DS3. Of the 33 attributes in this dataset, 12 were filtered by the rules of the classifier, whereas 21 were not selected by the same rules. In this case, note that a significant number of attributes was not chosen by the rules of the classifier. This means that the classifier was able to filter the most relevant features (12 features, see Appendix A.3) for the diseases represented by DS3, whereas the remaining features can be removed from the analysis, since they do not provide valuable information.

By analyzing the attributes in this dataset, we have that the FIBROSIS, AGE, ITCHING, and SPONGIO attributes have the greatest frequency of occurrence. In particular, FIBROSIS appears in 50% of the classes of this dataset (classes: psoriasis, seboreic dermatitis and cronic dermatitis), whereas AGE appears in 80% of rules in class-3 (lichen planus), ITCHING, and SPONGIO appear in 60% of rules in the same class-3. In particular, patients in each class are governed by the following relationships:

Class-0: patients holding $\{FIBROSIS = 0, SPONGIO = 0, ELONGATION > 0\}$;
Class-1: patients holding $\{FIBROSIS = 0, AGE = 20, DBORDERS \leq 2\}$;
Class-2: patients holding $\{BANDLIKE > 1, THINNING \neq 1\}$;
Class-3: this class supports four age-related subgroups of patients, namely, $\{AGE \geq 18, ITCHING \leq 1\}$, $\{AGE = 27, ITCHING < 2, SPONGIO > 0\}$, $\{AGE = 36, ITCHING \leq 1, SPONGIO > 0\}$ and $\{AGE = 62, SPONGIO > 0\}$;
Class-4: patients holding $\{FIBROSIS > 0, POLYPAPULES = 0\}$;
Class-5: patients holding $\{PERIFOLLI > 0, FOLLIPAPULES > 0\}$.

Note that, unlike the AGE feature, a value zero for the remaining features means that such a feature is not present in the patient, whereas a value greater than zero means that the patient presents the feature to a degree associated with the value. Consequently, with the results above, we can say that the study of these attributes can contribute to gain more insight about the diseases involved in such a dataset.

6. Conclusions

This work has proposed a machine learning method focused on genetic programming to render rule-based classifiers. Hence, this proposal has been aimed at inducing sets of logical rules able to learn the structure of the classes given in a dataset. We have applied the proposal to three clinical datasets (our concerning domain) and compared with other methods. In addition, we have identified the most reliable mutation operators regarding each dataset and, in that way, to improve the efficiency of our proposal. The results reached have been very promising when compared with other approaches. This proves the reliability of this approach to be used in the analysis of clinical data, which is our target data domain. Finally, we have disclosed certain relevant features from the logical rules found for each dataset involved in the experiment. Thereby, the proposal presented in this work can also be useful in the process of feature selection, since the attributes appearing in the rules of a classifier are the most important and so they discriminate the rest of attributes of the dataset. Related to the above, we have given an interpretation of the data by analyzing the dataset structures and the features of the rules found for each dataset. This prior knowledge can help the expert to establish a starting point for the study of the disease represented in the datasets.

Author Contributions: Conceptualization, methodology, validation, formal analysis, J.A.C.-G., E.C., J.L.J.S., and S.M.L.G.; investigation and writing—original draft preparation, J.A.C.-G. and Y.M.M.; writing—review and editing and supervision, E.C., J.L.J.S., and S.M.L.G.; resources, project administration, funding acquisition, S.M.L.G. All authors have read and agreed to the published version of the manuscript.

Funding: This research was funded by the MINISTERIO DE CIENCIA E INNOVACIÓN, Project: La desigualdad económica en la España contemporánea y sus efectos en los mercados, las empresas y el acceso a los recursos naturales y la tierra, Grant No. HAR2016-75010-R, corresponding to the research of Santiago M. López G.

Acknowledgments: This research has been supported by project "Intelligent and sustainable mobility supported by multi-agent systems and edge computing (InEDGEMobility): Towards Sustainable Intelligent Mobility: Blockchain-based framework for IoT Security", Reference: RTI2018-095390-B-C32, financed by the Spanish Ministry of Science, Innovation and Universities (MCIU), the State Research Agency (AEI) and the European Regional Development Fund (FEDER). The initial part of this research was also supported iCIS project (CENTRO-07-ST24-FEDER-002003), which has been co-financed by QREN, in the scope of the Mais Centro Program and European Union's FEDER.

Conflicts of Interest: The authors declare no conflict of interest.

Appendix A. Rule Based-Classifiers Rendered by the Evolutionary Method

Appendix A.1. Rules of the DS1 Dataset (Heart Dataset)

```
Number of rules in the Classifier: 7
number of attributes: 22
Number of classes: 2


(Class #0)
--------------------
IF (F13<0.37 ^ F16<=0.00 ^ F11=0.00) THEN Class := 0
IF (F5=1.00 ^ F21<=0.00 ^ F20<>1.00 ^ F10<1.00) THEN Class := 0
IF (F7<0.05 ^ F5<=0.72 ^ F17<=0.39 ^ F22<>1.00 ^ F4=0.00 ^ F6<>0.00) THEN
Class := 0
IF (F17<0.38 ^ F14<>0.00 ^ F3>=0.07 ^ F20<=0.00 ^ F15=0.00 ^ F21=1.00) THEN
Class := 0
IF (F13>=1.00 ^ F22<=0.68 ^ F1<1.00 ^ F14<=0.02 ^ F8<=0.00 ^ F11<>1.00) THEN
Class := 0
IF (F5<0.34 ^ F10<>0.00 ^ F21>=1.00 ^ F19=0.00 ^ F3=1.00 ^ F22<>1.00) THEN
Class := 0


(Class #1)
--------------------
IF (F8<=1.00) THEN Class := 1
```

Appendix A.2. Rules of the DS2 Dataset (Hepatitis Dataset)

```
Number of rules in the Classifier: 5
number of attributes: 19
Number of classes: 2

(Class #0)
--------------------
IF (FATIGUE<=1.00 ^ SEX=1.00 ^ ALBUMIN<=3.99 ^ PROTIME<=50.00 ^ PROTIME>28.85)
THEN Class := 0
IF (SPIDERS<=1.00 ^ SPLEEN>1.11 ^ ALBUMIN<>3.80 ^ AGE>=37.00 ^
ANTIVIRALS>=1.87 ^ ALK>=64.88 ^ ALK<95.00 ^ BILIRUBIN<>0.80) THEN
Class := 0
IF (ALK>104.77 ^ PROTIME>=56.16 ^ SGOT<=64.00 ^ ASCITES>=2.00 ^ ALK<>50.00 ^
```

```
ALBUMIN>=3.50 ^ AGE>=30.00) THEN Class := 0
```

```
(Class #1)
--------------------
IF (ALBUMIN>=2.90) THEN Class := 1
IF (SEX>1.03) THEN Class := 1
```

Appendix A.3. Rules of the DS3 Dataset (Dermatology Dataset)

```
Number of rules in the Classifier: 11
number of attributes: 34
Number of classes: 6
```

```
(Class #0)
--------------------
IF (fibrosis<=0.00 ^ elongation<>0.00 ^ spongio<=0.00) THEN Class := 0
```

```
(Class #1)
--------------------
IF (kphenom=0.00 ^ vacuoli<=0.97 ^ clubbing<=0.48 ^ follipapules<1.83 ^
fibrosis=0.00 ^ disappear<0.32 ^ thinning<=1.00) THEN Class := 1
IF (age=20.00 ^ dborders<2.00) THEN Class := 1
```

```
(Class #2)
--------------------
IF (bandlike>1.00 ^ thinning<>1.00) THEN Class := 2
```

```
(Class #3)
--------------------
IF (bandlike<=0.00 ^ PNL<3.00 ^ kphenom>0.00 ^ elongation<=0.00) THEN
Class := 3
IF (elongation<=0.00 ^ age>=18.00 ^ itching<1.11 ^ disappear<>0.00) THEN
Class := 3
IF (itching<2.00 ^ inflam=2.00 ^ age=27.00 ^ spongio>0.00) THEN Class := 3
IF (age=36.00 ^ spongio>0.00 ^ inflam=2.00 ^ itching<=1.00) THEN Class := 3
IF (age=62.00 ^ spongio<>0.00 ^ sawtooth=0.00) THEN Class := 3
```

```
(Class #4)
--------------------
IF (fibrosis>0.00 ^ polypapules<=0.00) THEN Class := 4
```

```
(Class #5)
--------------------
IF (perifolli>0.00 ^ follipapules<>0.00) THEN Class := 5
```

Appendix B. Test Charts of the Mutation Operators

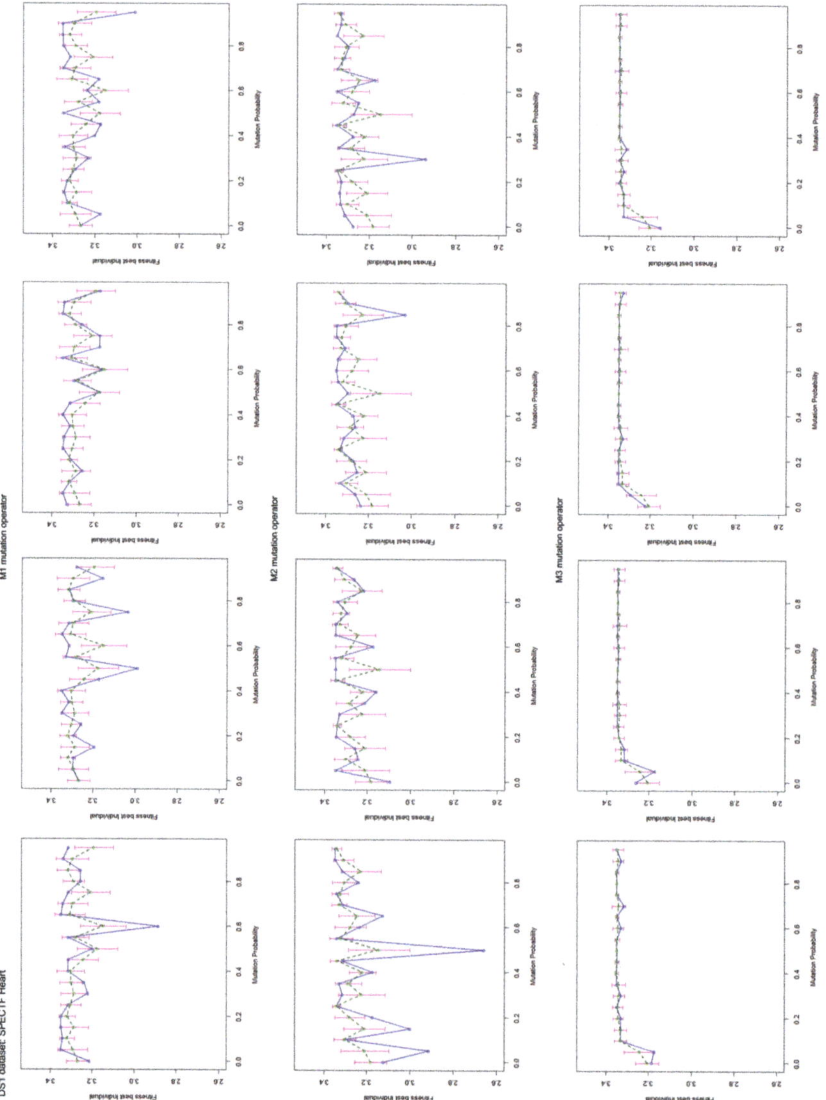

Figure A1. Mutation tests for mutation operators M1, M2, and M3 for the DS1 dataset. Each row (with four graphics) in the figure corresponds to the same mutation operator and each graphic corresponds to 20 executions of the evolutionary method for 20 mutation probability values with step 0.05. The blue curve represents fitness values against mutation values. The green curve represents the mean fitness values from the four graphics in the same row and the pink lines state the standard error bars.

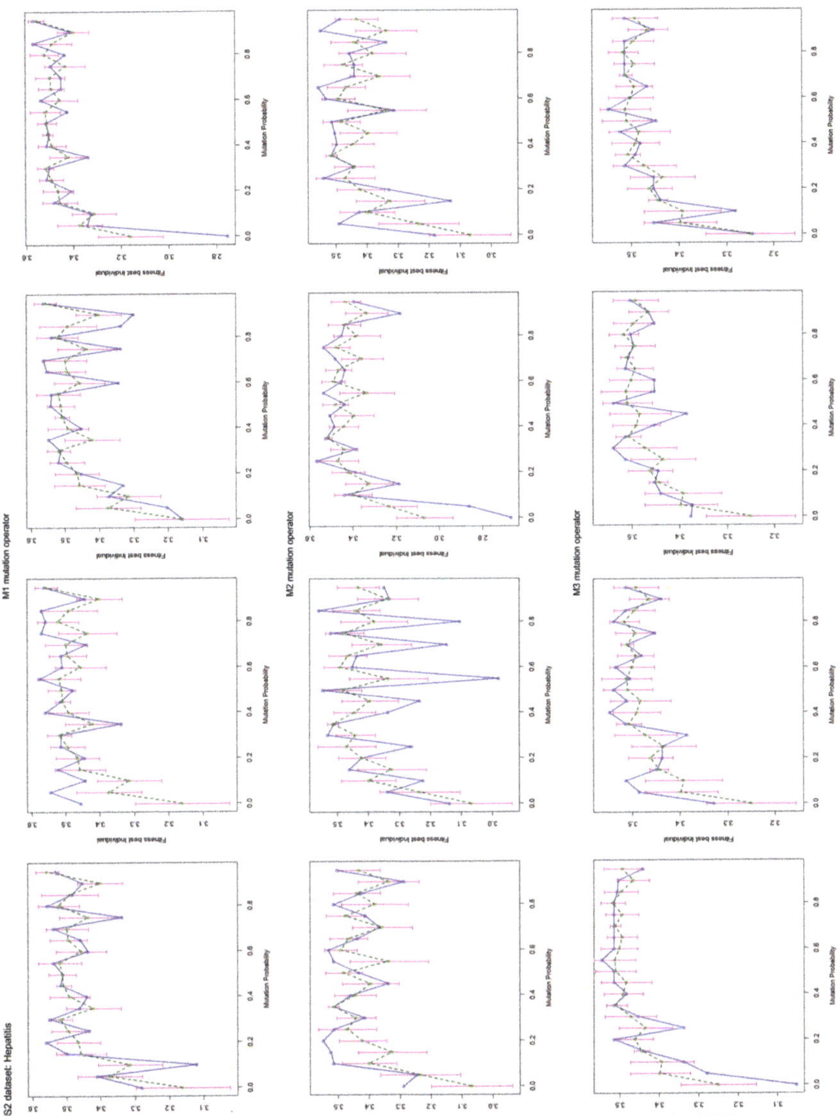

Figure A2. Mutation tests for mutation operators M1, M2, and M3 for the DS2 dataset. Each row of four graphics in the figure corresponds to the same mutation operator and each graphic corresponds to 20 executions of the evolutionary method for 20 mutation probability values with step 0.05. The blue line represents each fitness value for each mutation value. The green line represents the mean values from the four graphics in the same row and pink lines state the standard error bars.

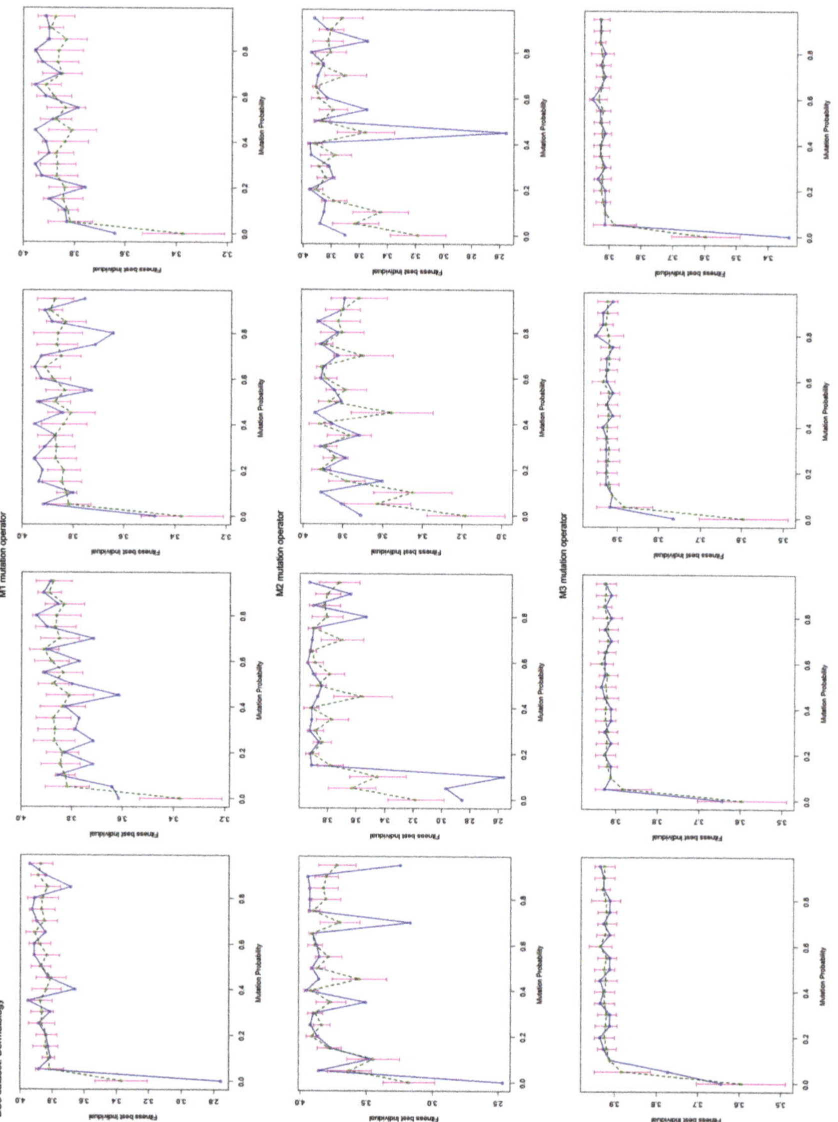

Figure A3. Mutation tests for mutation operators M1, M2, and M3 for the DS3 dataset. Each row of four graphics in the figure corresponds to the same mutation operator and each graphic corresponds to 20 executions of the evolutionary method for 20 mutation probability values with step 0.05. The blue line represents each fitness value for each mutation value. The green line represents the mean values from the four graphics in the same row and pink lines state the standard error bars.

References

1. Bandyopadhyay, S.; Pal, S.K. *Classification and Learning Using Genetic Algorithms: Applications in Bioinformatics and Web Intelligence*; Natural Computing Series; Springer: Berlin/Heidelberg, Germany, 2007.
2. Bonelli, P.; Parodi, A. An efficient classifier system and its experimental comparison with two representative learning methods on three medical domains. In Proceedings of the 4th International Conference Genetic Algorithms (ICGA), San Diego, CA, USA, July 1991; pp. 288–295.
3. Hong, J.H.; Cho, S.B. The classification of cancer based on DNA microarray data that uses diverse ensemble genetic programming. *Artif. Intell. Med.* **2006**, *36*, 43–58. [CrossRef] [PubMed]
4. Kumar, T.P.; Iba, H. Prediction of Cancer Class with Majority Voting Genetic Programming Classifier Using Gene Expression Data. *IEEE/ACM Trans. Comput. Biol. Bioinform.* **2009**, *6*, 353–367.
5. Kumar, R.; Verma, R. Classification Rule Discovery for Diabetes Patients by Using Genetic Programming. *Int. J. Soft Comput. Eng. IJSCE* **2012**, *2*, 183–185.
6. Larranaga, P.; Calvo, B.; Santana, R.; Bielza, C.; Galdiano, J.; Inza, I.; Lozano, J.A.; Armañanzas, R.; Santafé, G.; Pérez, A.; et al. Machine learning in bioinformatics. *Briefings Bioinform.* **2006**, *7*, 86–112.
7. Liu, K.H.; Xu, C.G. A genetic programming-based approach to the classification of multiclass microarray datasets. *Bioinformatics* **2009**, *25*, 331–337. [CrossRef] [PubMed]
8. Maulik, U.; Bandyopadhyay, S.; Mukhopadhyay, A. *Multiobjective Genetic Algorithms for Clustering: Applications in Data Mining and Bioinformatics*; Springer: Berlin/Heidelberg, Germany, 2011.
9. na Reyes, C.A.P.; Sipper, M. Evolutionary computation in medicine: An overview. *Artif. Intell. Med.* **2000**, *19*, 1–23. [CrossRef]
10. Podgorelec, V.; Kokol, P.; Stiglic, M.M.; Hericko, M.; Rozrnan, I. Knowledge discovery with classification rules in a cardiovascular dataset. *Comput. Methods Programs Biomed.* **2005**, *1*, 539–549. [CrossRef]
11. Soni, J.; Ansari, U.; Sharma, D.; Soni, S. Intelligent and Effective Heart Disease Prediction System using Weighted Associative Classifiers. *Int. J. Comput. Sci. Eng. IJCSE* **2011**, *3*, 2385–2392.
12. Tsakonas, A.; Dounias, G.; Jantzen, J.; Axer, H.; Bjerregaard, B.; von Keyserlingk, D.G. Evolving rule-based systems in two medical domains using genetic programming. *Artif. Intell. Med.* **2004**, *32*, 195–216. [CrossRef]
13. Vargas, C.M.B.; Chidambaram, C.; Hembecker, F.; Silvério, H.L. *Computational Biology and Applied Bioinformatics*; Chapter A Comparative Study of Machine Learning and Evolutionary Computation Approaches for Protein Secondary Structure Classification; InTech: London, UK, 2011; pp. 239–258.
14. Wolberg, W.H.; Mangasarian, O.L. Multisurface method of pattern separation for medical diagnosis applied to breast cytology. *Proc. Natl. Acad. Sci. USA* **1990**, *87*, 9193–9196. [CrossRef]
15. Lucas, P. Analysis of notions of diagnosis. *Artif. Intell.* **1998**, *12*, 295–343. [CrossRef]
16. Lucas, P. Prognostic methods in medicine. *Artif. Intell.* **1999**, *15*, 105–119. [CrossRef]
17. Dong, X.; Yu, Z.; Cao, W.; Shi, Y.; Ma, Q. A survey on ensemble learning. In *Frontiers of Computer Science*; Springer: Berlin/Heidelberg, Germany, 2019; [CrossRef]
18. Ramos, J.; Castellanos-Garzón, J.A.; González-Briones, A.; de Paz, J.F.; Corchado, J.M. An agent-based clustering approach for gene selection in gene expression microarray. In *Interdisciplinary Sciences: Computational Life Sciences*; Springer: Berlin/Heidelberg, Germany, 2017; Volume 9, pp. 1–13.
19. Castellanos-Garzón, J.A.; Ramos, J.; González-Briones, A.; de Paz, J.F. A Clustering-Based Method for Gene Selection to Classify Tissue Samples in Lung Cancer. In *10th International Conference on Practical Applications of Computational Biology & Bioinformatics, Advances in Intelligent Systems and Computing*; Saberi Mohamad, M., Rocha, M., Fdez-Riverola, F., Domínguez Mayo, F., De Paz, J., Eds.; Springer: Cham, Switzerland, 2016; Volume 477, pp. 99–107.
20. Castellanos-Garzón, J.A.; Ramos, J. A Gene Selection Approach based on Clustering for Classification Tasks in Colon Cancer. *ADCAIJ Adv. Distrib. Comput. Artif. Intell. J.* **2015**, *4*, 1–10. [CrossRef]
21. González-Briones, A.; Ramos, J.; De Paz, J.F. A drug identification system for intoxicated drivers based on a systematic review. *ADCAIJ Adv. Distrib. Comput. Artif. Intell. J.* **2015**, *4*, 83–101.

22. Pappa, G.L.; Freitas, A.A. Evolving rule induction algorithms with multi-objective grammar-based genetic programming. In *Knowledge and Information Systems*; Springer: Berlin/Heidelberg, Germany, 2009; Volume 19, pp. 283–309.

23. Alcalá-Fdez, J.; Sánchez, L.; García, S.; delJesus, M.J.; Ventura, S.; Garrell, J.M.; Otero, J.; Romero, C.; Bacardit, J.; Rivas, V.M.; et al. KEEL: A software tool to assess evolutionary algorithms for data mining problems. In*Soft Computing*; Springer: Berlin/Heidelberg, Germany, 2009; Volume 13, pp. 307–318.

24. Fernández, A.; García, S.; Luengo, J.; Bernadó-Mansilla, E.; Herrera, F. Genetics-Based Machine Learning for Rule Induction: State of the Art, Taxonomy, and Comparative Study. *IEEE Trans. Evol. Comput.* **2010**, *14*, 913–941. [CrossRef]

25. Oyebode, O.K.; Adeyemo, J.A. Genetic Programming: Principles, Applications and Opportunities for Hydrological Modelling. *World Acad. Sci. Eng. Technol. Int. J. Environ. Ecol. Geol. Min. Eng.* **2014**, *8*, 310–316.

26. Ghaheri, A.; Shoar, S.; Naderan, M.; Hoseini, S.S. The applications of genetic algorithms in medicine. *Oman Med. J.* **2015**, *30*, 406–416. [CrossRef]

27. Karnan, M.; Thangavel, K. Automatic detection of the breast border and nipple position on digital mammograms using genetic algorithm for asymmetry approach to detection of microcalcifications. In *Computer Methods and Programs in Biomedicine*; Elsevier: Amsterdam, The Netherlands, 2007; Volume 87, pp. 12–20.

28. Gudmundsson, M.; El-Kwae, E.A.; Kabuka, M.R. Edge detection in medical images using a genetic algorithm. *IEEE Trans. Med Imaging* **1998**, *17*, 469–474. [CrossRef]

29. Bhandarkar, S.M.; Zhang, Y.; Potter, W.D. An edge detection technique using genetic algorithm-based optimization. *Pattern Recognit.* **1994**, *27*, 1159–1180. [CrossRef]

30. Jiang, J.; Yao, B.; Wason, A.M. A genetic algorithm design for microcalcification detection and classification in digital mammograms. In *Computerized Medical Imaging and Graphics*; Elsevier: Amsterdam, The Netherlands, 2007; Volume 31, pp. 49–61.

31. Yao, B.; Jiang, J.; Peng, Y. A CBR driven genetic algorithm for microcalcification cluster detection. In *International Conference on Knowledge Engineering and Knowledge Management*; Springer: Berlin/Heidelberg, Germany, 2004; pp. 494–496.

32. Bevilacqua, A.; Campanini, R.; Lanconelli, N. A distributed genetic algorithm for parameters optimization to detect microcalcifications in digital mammograms. In *Workshops on Applications of Evolutionary Computation*; Springer: Berlin/Heidelberg, Germany, 2001; pp. 278–287.

33. Baum, K.G.; Schmidt, E.; Rafferty, K.; Król, A.; Helguera, A. Evaluation of novel genetic algorithm generated schemes for positron emission tomography (PET)/magnetic resonance imaging (MRI) image fusion. *J. Digit. Imaging* **2011**, *24*, 1031–1043. [CrossRef]

34. de Carvalho Filho, A.O.; de Sampaio, W.B.; Silva, A.C.; de Paiva, A.C.; Nunes, R.A.; Gattass, M. Automatic detection of solitary lung nodules using quality threshold clustering, genetic algorithm and diversity index. *Artif. Intell. Med.* **2014**, *60*, 165–177. [CrossRef] [PubMed]

35. Asuntha, A.; Singh, N.; Srinivasan, A. PSO, Genetic Optimization and SVM Algorithm used for Lung Cancer Detection. *J. Chem. Pharm. Res.* **2016**, *8*, 351–359.

36. Alshamlan, H.M.; Badr, G.H.; Alohali, Y.A. Genetic Bee Colony (GBC) algorithm: A new gene selection method for microarray cancer classification. *Comput. Biol. Chem.* **2015**, *56*, 49–60. [CrossRef] [PubMed]

37. Latkowski, T.; Osowski, S. Computerized system for recognition of autism on the basis of gene expression microarray data. *Comput. Biol. Med.* **2015**, *56*, 82–88. [CrossRef] [PubMed]

38. Arabasadi, Z.; Alizadehsani, R.; Roshanzamir, M.; Moosaei, H.; Yarifard, A.A. Computer aided decision-making for heart disease detection using hybrid neural network-Genetic algorithm. *Comput. Methods Programs Biomed.* **2017**, *141*, 19–26, [CrossRef] [PubMed]

39. Li, H.; Yuan, D.; Ma, X.; Cui, D.; Cao, L. *Genetic algorithm for the Optimization of Features and Neural Networks in ECG Signals Classification*; Resreport 7, Scientific Reports; Springer Nature: Berlin/Heidelberg, Germany, 2017.

40. Lin, T.; Huang, Y.; Lin, J.I.; Balas, V.E.; Srinivasan, S. Genetic algorithm-based interval type-2 fuzzy model identification for people with type-1 diabetes. In Proceedings of the IEEE International Conference on Fuzzy Systems (FUZZ-IEEE), Naples, Italy, 9–12 July 2017; pp. 1–6. [CrossRef]

41. Nguyen, L.B.; Nguyen, A.V.; Ling, S.H.; Nguyen, H.T. Combining genetic algorithm and Levenberg-Marquardt algorithm in training neural network for hypoglycemia detection using EEG signals. In Proceedings of the Engineering in Medicine and Biology Society (EMBC), 2013 35th Annual International Conference of the IEEE, Osaka, Japan, 3–7 July 2013; pp. 5386–5389.

42. Ocak, H. A medical decision support system based on support vector machines and the genetic algorithm for the evaluation of fetal well-being. *J. Med Syst.* **2013**, *37*, 9913. [CrossRef] [PubMed]

43. Nyathi, T.; Pillay, N. Automated Design of Genetic Programming Classification Algorithms Using a Genetic Algorithm. In *Applications of Evolutionary Computation: 20th European Conference, EvoApplications 2017, Proceedings of the Part II, Amsterdam, The Netherlands, 19–21 April 2017*; Squillero, G., Sim, K., Eds.; Chapter Applications of Evolutionary Computation. EvoApplications 2017. Lecture Notes in Computer Science; Springer International Publishing: Cham, Switzerland, 2017; Volume 10200, pp. 224–239. [CrossRef]

44. Naredo, E.; Trujillo, L.; Legrand, P.; Silva, S.; Muñoz, L. Evolving genetic programming classifiers with novelty search. *Inf. Sci.* **2016**, *369*, 347–367. [CrossRef]

45. Dick, G., Improving geometric semantic genetic programming with safe tree initialisation. In *Genetic Programming: 18th European Conference, EuroGP 2015, Copenhagen, Denmark, 8–10 April 2015*; Machado, P., Heywood, M.I., McDermott, J., Castelli, M., García-Sánchez, P., Burelli, P., Risi, S., Sim, K., Eds.; Chapter European Conference on Genetic Programming, EuroGP 2015: Genetic Programming; Springer International Publishing: Cham, Switzerland, 2015; pp. 28–40, [CrossRef]

46. Alotaiby, T.N.; Alrshoud, S.R.; Alshebeili, S.A.; Alhumaid, M.H.; Alsabhan, W.M. Epileptic MEG Spike Detection Using Statistical Features and Genetic Programming with KNN. *J. Healthc. Eng. Hindawi* **2017**, *2017*, 7. [CrossRef]

47. Wang, C.S.; Juan, C.J.; Lin, T.Y.; Yeh, C.C.; Chiang, S.Y. Prediction Model of Cervical Spine Disease Established by Genetic Programming. In Proceedings of the 4th Multidisciplinary International Social Networks Conference (MISNC '17); ACM: New York, NY, USA, July 2017; pp. 381–386, [CrossRef]

48. Burks, A.R.; Punch, W.F. Genetic Programming for Tuberculosis Screening from Raw X-ray Images. In Proceedings of the Genetic and Evolutionary Computation Conference (GECCO '18), New York, NY, USA, 15–19 July 2018; pp. 1214–1221, [CrossRef]

49. Tan, P.N.; Steinbach, M.; Kumar, V. *Introduction to Data Mining*; Addison-Wesley: Boston, MA, USA, 2006.

50. Freitas, A.A. *Soft Computing for Knowledge Discovery and Data Mining, Part II*; Chapter A Review of Evolutionary Algorithms for Data Mining; Springer: Berlin/Heidelberg, Germany, 2008; pp. 79–111.

51. Witten, I.H.; Frank, E. *Data Mining: Practical Machine Learning Tools and Techniques*, 2nd ed.; Morgan Kaufmann: Burlington, MA, USA, 2005.

52. Pappa, G.L.; Freitas, A.A. *Automating the Design of Data Mining Algorithms: An Evolutionary Computation Approach*; Springer: Berlin/Heidelberg, Germany, 2010.

53. Espejo, P.G.; Ventura, S.; Herrera, F. A Survey on the Application of Genetic Programming to Classification. *IEEE Trans. Syst. Man Cybern. Part Appl. Rev.* **2010**, *40*, 121–144. [CrossRef]

54. Freitas, A.A. A survey of evolutionary algorithms for data mining and knowledge discovery. Advances in Evolutionary Computation. In *Advances in Evolutionary Computation*; Ghosh, A., Tsutsui, S., Eds.; Springer: Berlin/Heidelberg, Germany, 2002; pp. 819–845.

55. Flach, P. *MACHINE LEARNING: The Art and Science of Algorithms that Make Sense of Data*; Cambridge University Press: Cambridge, UK, 2012.

56. Witten, I.H.; Frank, E.; Hall, M.A. *Data Mining: Practical Machine Learning, Tools and Techniques*, 3rd ed.; Elsevier Inc.: Amsterdam, The Netherlands, 2011.

57. Goldberg, D.E. *Genetic Algorithms in Search, Optimization and Machine Learning*; Addison-Wesley: Reading, MA, USA, 1989.

58. Bacardit, J.; Goldberg, D.E.; Butz, M.V. Improving the performance of a Pittsburgh learning classifier system using a default rule. In *Proceedings Revised Select Papers International Workshop Learning Classifier Systems*; Springer: Berlin/Heidelberg, Germany, 2007; pp. 291–307.

59. Haupt, R.L.; Haupt, S.E. *Practical Genetic Algorithms*, 2nd, ed.; John Wiley & Sons Inc.: Hoboken, NJ, USA, 2004.

60. Castellanos-Garzón, J.A.; Díaz, F. An Evolutionary Computational Model Applied to Cluster Analysis of DNA Microarray Data. *Expert Syst. Appl.* **2013**, *40*, 2575–2591. [CrossRef]

61. R Core Team. *R: A Language and Environment for Statistical Computing*; R Foundation for Statistical Computing: Vienna, Austria, 2018.

62. Blake, C.; Merz, C. *Repository of Machine Learning Databases (UCI)*; Center for Machine Learning and Intelligent Systems: Irvine, CA, USA; 1998.

63. Wu, X.; Kumar, V.; Ross Quinlan, J.; Ghosh, J.; Yang, Q.; Motoda, H.; McLachlan, G.; Ng, A.; Liu, B.; Yu, P.; et al. Top 10 algorithms in data mining. *Knowl. Inf. Syst.* **2008**, *14*, 1–37, [CrossRef]

64. Moerland, P. *Mixture for Latent Variable Models for Density Estimation and Classification*; Technical Report; Dalle Molle Institution for Perceptual Artificial Intelligencie, IDIAP: Martigny, Switzerland, 2000.

MDPI

St. Alban-Anlage 66

4052 Basel

Switzerland

Tel. +41 61 683 77 34

Fax +41 61 302 89 18

www.mdpi.com

Processes Editorial Office

E-mail: processes@mdpi.com

www.mdpi.com/journal/processes